"STEM In Music"

Principles of Engineering
Teaching, and Learning Concepts
Educational Curriculum / Program

By:

Michael A. Johnson

Exploring, Learning, and Discovering endless Possibilities.

STEM Educational Enrichment Services LLC
STEM In Music

My Music Instrument (m2i)

By:

Mr. Michael A. Johnson

STEM Educational Enrichment Services LLC.

"STEM In Music"

40 years ago

Michael A. Johnson
Murrell Dobbins AVTS
High School Graduation
1978

Bradley Guitar - Les Paul Style

The Same Guitar

40 years Later

Michael A. Johnson
TMS STEM Studios
"STEM In Music"
Guitar Renovation Lab

"STEM In Music"

CEO / Author / Founder:

Michael A. Johnson

Michael A. Johnson is a retired educator from the state of Pennsylvania, Author of "What's In Me", STEM Consultant, Motivational Speaker, Curriculum Designer of "STEM In Music", Inventor of My Music Instrument (m2i), and Owner of STEM Educational Enrichment Services LLC.

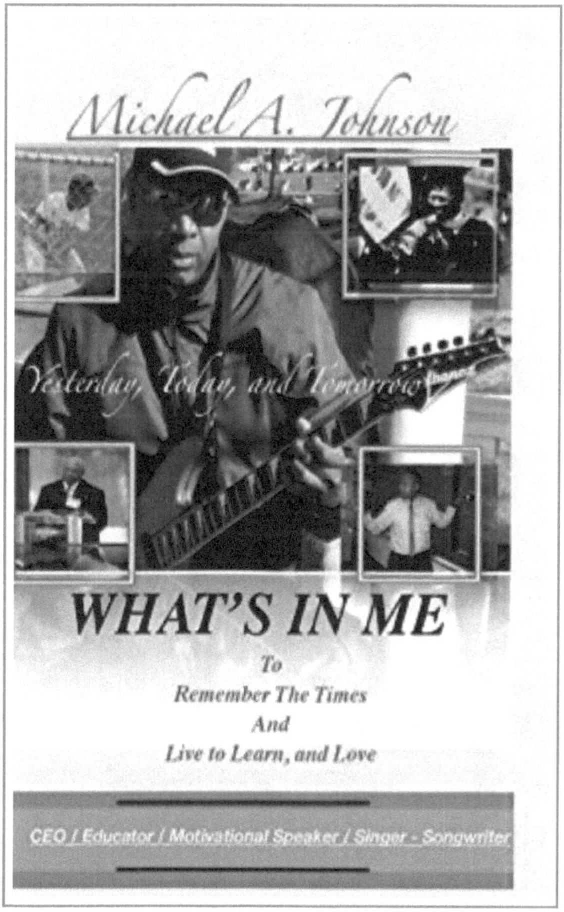

Mr. Johnson has thirty-two years experience teaching K through Post Graduate School, forty-seven years as a singer / songwriter / musician, and nineteen years as an educational researcher.

STEM Educational Enrichment Services LLC.

"STEM In Music"

By:

Michael A. Johnson

STEM Educational Enrichment Services LLC founded September 2009 with a goal introducing key concepts in Principles of Engineering Teaching, and Learning Concepts. Technological areas, and Music concepts include: ITEEA (International Technology Engineering Educators Association), TEEAP (Technology Engineering Education Association of Pennsylvania), Little Kids Rock / Modern Band Summit, and now *"STEM In Music".*

STEM In Music is an educational enrichment curriculum / program written, designed, produced, and create by Michael A. Johnson Fall Semester (September) 2018.

The purpose implementing STEM In Music provides creative learning skills.

Music conveys an essential role in manifesting, and preparing students with 21st century global career opportunities. Today, STEM revolutionizes how we learn, listen, and play music. It branches reach, and enter many needed careers; especially Music *(STEM IN Music)*.

STEM Educational Enrichment Services LLC Principles of Engineering "STEM In Music" covers Research, Investigation, Technological Literacy, Systems, Resources, Tools, Design / Engineering Process, Computer Aided Design, Manufacturing (3D Modeling / Printing), Laws of Physics, Measurements, and Mathematical Conversions.

STEM Educational Enrichment Services LLC STEM In Music activities include: ***Guitar Renovations*, *Guitar Notation*, *Song Writing, Recording, Producing***, and ***Musical Instrument Invention***.

It's an honor presenting
STEM Educational Enrichment
Services LLC
<u>"STEM In Music" curriculum / program.</u>

This curriculum / program enhances schools, businesses, communities, and industries Science, Technology, Engineering, and Mathematical learning concepts.

STEM In Music

Purpose / Passion / Plan

The purpose, passion, and plan implementing "STEM In Music" key focus integrates academic studies with natural creative abilities. STEM (Science, Technology, Engineering, and Mathematics) offers multiple documentation applications, and scientific learning activities.

STEM In Music documentation applications include: Technological Literacy, Systems, and Resources.

STEM In Music Scientific Learning Process includes: Possibility Statement, Design Brief, Getting to Know The Possibilities, Exploring Concepts, Possible Solutions, Testing, and Evaluations.

"STEM In Music"

Goal

"STEM In Music goal provides historical, current, and future developments understanding Science, Technology, Engineering, and Mathematics in Music using ITEEA (International Technology and Engineering Educators Association), TEEAP (Technology & Engineering, Education Association of Pennsylvania) and Little Kids Rock / Modern Band summit. Some Key areas include: The Nature of Technology, Technology and Society, Technology and Design, The use of Technology, and Abilities in a Technological World.

The Nature of Technology:

1. Technological Literacy
2. Technological Systems
3. Technological Resources
4. Technological Design Process

Technology and Society:

1. Safety
2. Musical Instrument Research & Development
3. Invention & Innovation Design Process
4. Power & Energy

Technology and Design:

1. Musical Instrument Design
2. Computer Aided Design
3. Guitar Renovation
4. 3D Modeling (Musical Instrument)

Technological World:

1. Principles of Engineering
2. Integrated Circuits & Systems

Abilities in a Technological World

1. Songwriting
2. Music Recording / Sound Engineering

STEM In Music Lab Activities:

Guitar Renovation Lab

The Nature of Technology
Technological Literacy
Technology Systems

Guitar Notation Lab

Abilities in a Technological World
Technological Resources

Song Writing Lab

Abilities in a Technological World
Technological Resources

Music Instrument Invention Lab I

Technology and Society
Music Research and Development
Safety

Music Instrument Invention Lab II

Technology and Design
Invention, and Innovation Design Process
CAD (Computer Aided Design)
3D Modeling / Printing (Makerbot)
MIT Makey-Makey integrated circuit
Technological Systems:
Input: (Play-Doh touch controls)
Process: (Electrical wires, Breadboard, Integrated Circuit, usb connector),
Output: (Apple Computer / Garageband Software)
Feedback: (Brass Fasteners Musical Notation Controls)

"STEM In Music" Musical Instrument Invention must be hand held.

STEM Educational Enrichment Services LLC The Nature of Technology:

Technological Literacy
Technological Systems
Technological Resources

One of the most powerful technological process is the ability to think.

"STEM In Music " Technological Literacy, Systems, and Resources defines clear understanding The Nature of Technology.

Technological Literacy helps accumulate information understanding a need. Technological Systems are designed controlling total operation exploring a need. Technological Resources provide a mean in exploring possibilities according to developing a need.

Guitar Renovation Lab

The Nature of Technology

Technological Literacy

<u>Design Brief</u> / <u>Problem Statement</u>

"STEM in Music" - The Nature of Technology lab projects include: Guitar Renovation, Guitar Musical Notation, Playing Guitar, Recording Music, and Songwriting.

Possibility Statement:

Guitar Renovation Lab needs a guitar to Repair, Play, Learn Music Notations, Write Songs, and Produce a Recording. The guitar has missing parts. You must use Technological Design Process during Lab Activity.

Bradley Guitar *(Les Paul Style)*

*This is my second guitar **(1976)***

Guitar Renovation Lab

The Nature of Technology

Technological Literacy

Design Brief

Find a use guitar in need of new parts, and identify each section.

STEM Educational Enrichment Services LLC. *"STEM In Music"* guitar renovation involves learning, and understanding Design, Measurements, Mechanisms, Materials, Systems.

Guitar Structural Design & Parts:

1. Head
2. Neck
3. Body
4. Frets
5. Bridge

Guitar Head:

The *guitar head* supports strings, and tuning Mechanisms. Standard guitar has six strings (E6, A, D, G, B, E1)

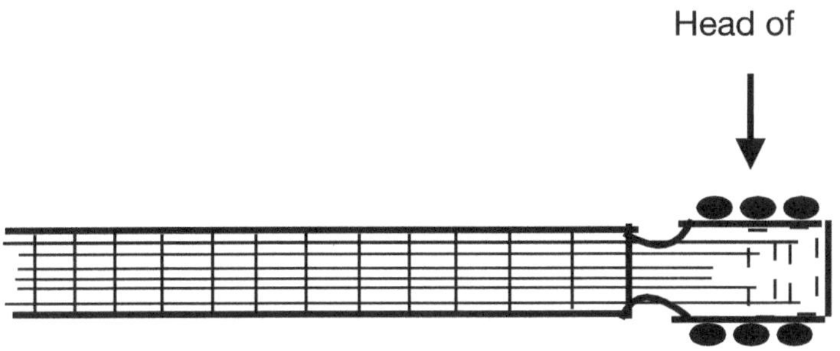

Guitar Renovation Lab

The Nature of Technology

Technological Systems

Research & Development

Guitar Head consist of tuning pegs & knobs. The neck of the guitar has a nut, frets, strings, and notes.

Guitar Head, and Neck Location & Purpose:

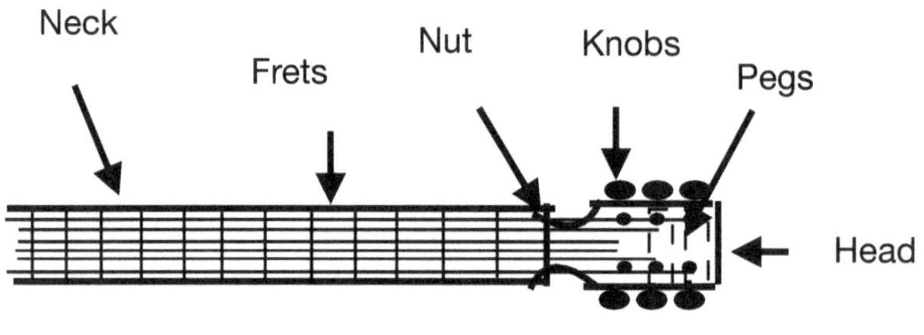

Guitar Head:

Tuning pegs support guitar strings.

Tuning knobs tighten, and losing strings for accurate tuning,

Neck:

Nut holds string in place.

Frets play a specific note.

Strings are used to create guitar notation.

Notes can be played individually, or collectively.

Guitar Head & Neck Materials:

Head of guitar is made out of: Wood

Tuning pegs are made out of: Metal, or Plastic

Tuning Knobs are made out of: Metal, Acrylic, or Plastic

Guitar Renovation Lab

The Nature of Technology

Technological Literacy

Research & Development

Guitar Head and Neck Mechanisms:

Tuning pegs (Lever Tuning knobs), Lever, (Gears and Screw), Nut (Wedge), Fret (Wedge) Strings (Lever), Notes on the Neck (Lever).

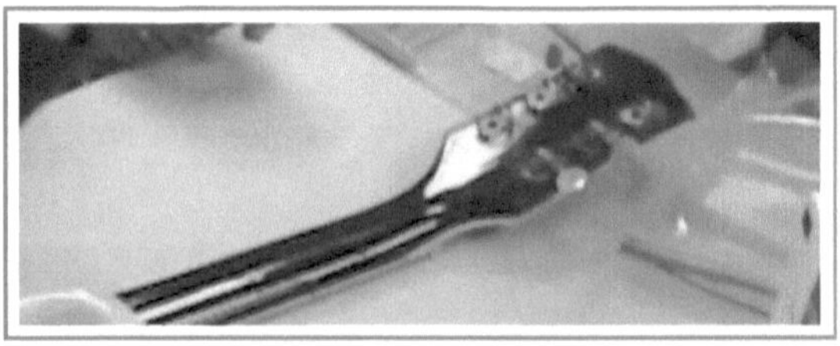

Guitar Renovation Lab

The Nature of Technology

Technological Systems

Guitar Body

The body on the guitar consists of:

Pick-ups

Bridge

Toggle Switch

Volume / Treble control knobs

Guitar Renovation Lab

The Nature of Technology

Technological Systems

Guitar Body

Guitar Body Location & Purpose:

Pick-ups are located on the body. They amplify sound from guitar strings.

Bridge is a devices supporting guitar strings.

Toggle switch controls pick-ups.

Volume / Treble knobs control the amount of volume, and treble.

Output Jack connects guitar with an amplifier.

Body of Guitar

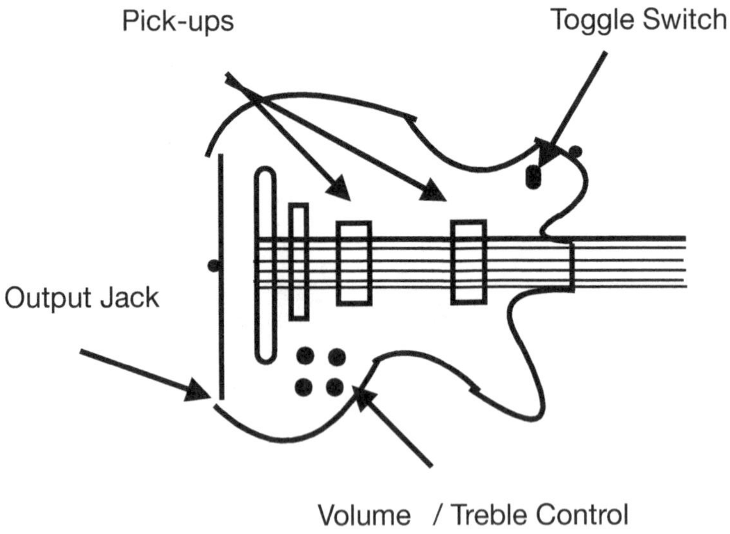

Guitar Body Materials:

Body is made out of: Wood, and Plastic.

Bridge is made out of: Wood, Plastic, or Metal.

Toggle switch is made out of: Plastic, and Metal.

Volume / Treble knobs is made out of: Acrylic, or Plastic.

Guitar Renovation Lab

The Nature of Technology

Technological Systems

Guitar Body

Guitar Body Mechanisms:

Pick-ups: Electrical System, Screw

Bridge: Wedge

Toggle Switch: Lever

Volume / Treble knobs: Screw

Fret: Wedge

Strings: Lever

Notes on the Neck: Lever

Guitar Notation Lab

Abilities in a Technological World

Technological Resourcses

Guitar Notation

Design Brief: Learn open ***Strings Notes***

Notes ***Strings***

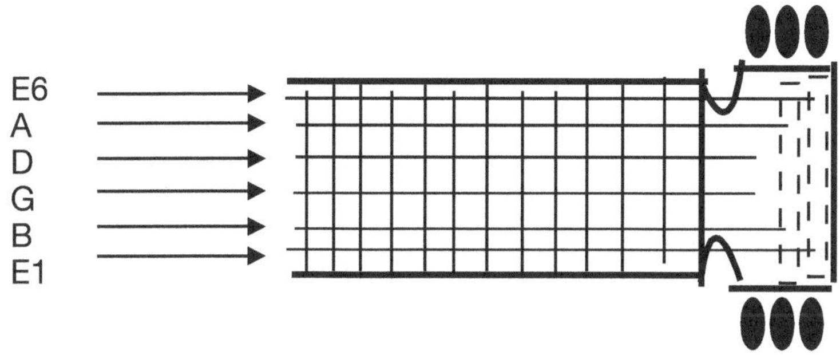

Design Brief: Learn ***Fret Positions***

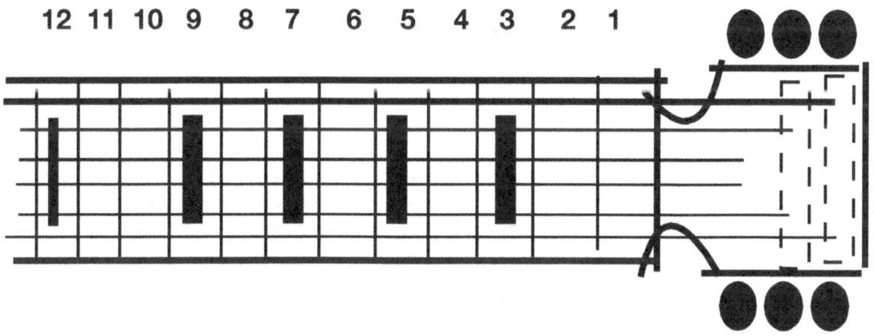

Guitar Notation Lab

The Nature of Technology

Technological Resourcses

Guitar Notation

Design Brief: Learn C major notes on **E1** String

(***Open String,*** ***1st Fret*** and ***3rd Fret***)

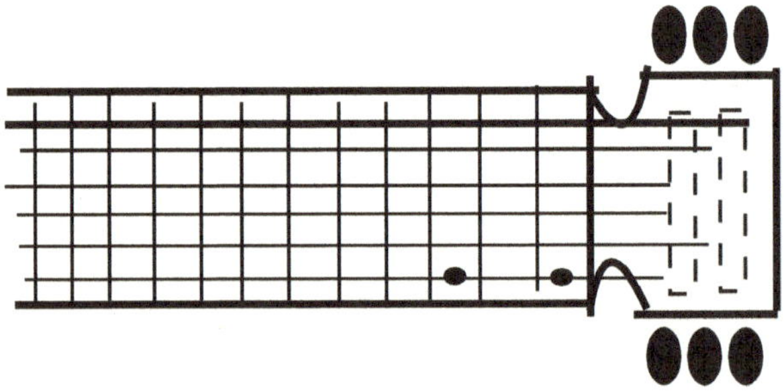

Design Brief: Learn C major notes on **B** String

Open String, ***1st Fret*** and ***3rd Fret***)

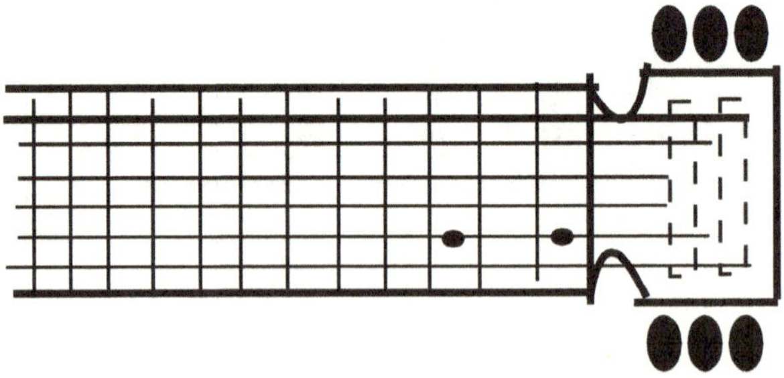

Design Brief: Learn C major notes on **G** String
(***Open String,*** and ***2nd Fret***)

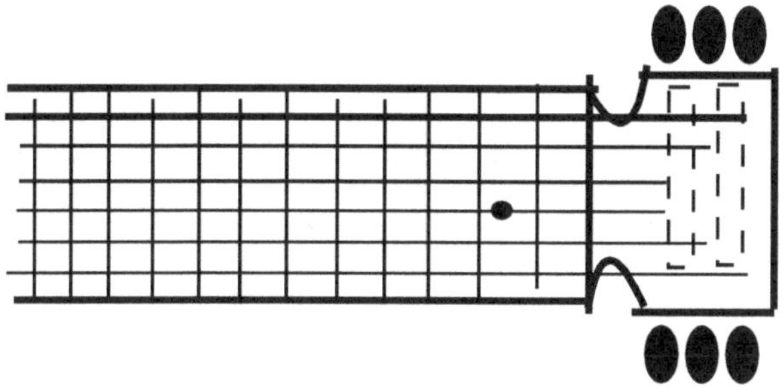

Design Brief: Learn C major notes on **D** String
(***Open String, 2nd Fret*** and ***3rd Fret***)

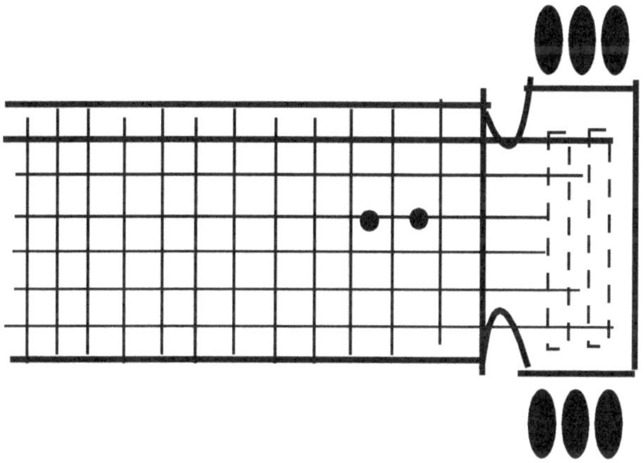

<u>Design Brief:</u> Learn C major notes on **A** String
(***<u>Open String,</u>*** ***<u>2nd Fret</u>*** and ***<u>3rd Fret</u>***

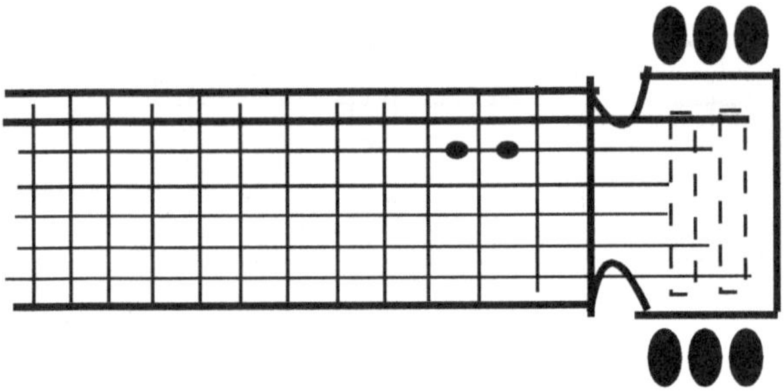

<u>Design Brief:</u> Learn C major notes (First position) on **E6** String.
(***<u>Open String,</u>*** ***<u>1st, and 3rd Fret</u>***)

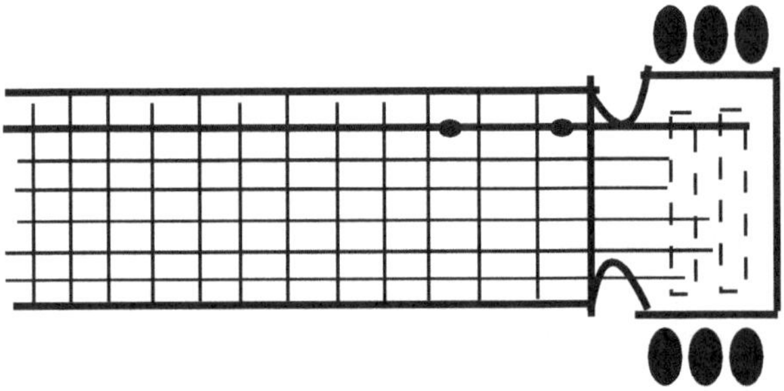

Guitar Notation Lab

The Nature of Technology

Technological Resourcses

Guitar Notation

Design Brief: Learn C major Scale (First position) on all **6** Strings.

E6: ***Open String, 1st,*** *and* ***3rd Fret***)
A: ***Open String, 2nd,*** *and* ***3rd Fret***)
D: ***Open String, 2nd,*** *and* ***3rd Fret***)
G: ***Open String, 2nd Fret***)
B: ***Open String, 1st,*** *and* ***3rd Fret***)
E1: ***Open String, 1st,*** *and* ***3rd Fret***)

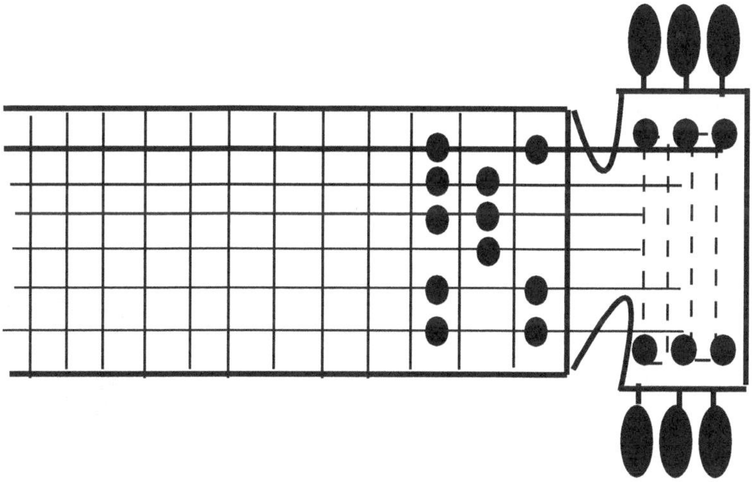

Guitar Notation Lab

The Nature of Technology
Technological Resourcses
Guitar Notation

Design Brief: Learn C major chord (First position).

C: ***A String, 3rd Fret***
E: ***D String, 2nd Fret***
G: ***Open String***
C: ***B String, 1st Fret***
E: ***Open String***

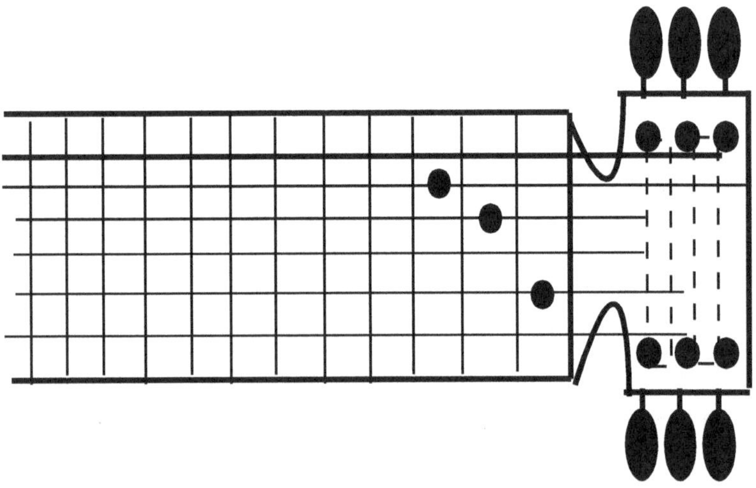

<u>Guitar Notation Lab</u>

The Nature of Technology

Technological Resourcses

Guitar Notation

<u>Design Brief:</u> Learn F major chord (First position).

F: ***<u>D String, 3rd Fret</u>***
A: ***<u>G String, 2nd Fret</u>***
C: ***<u>B String Bar 1st Fret</u>***
F: ***<u>E String Bar, 1st Fret</u>***

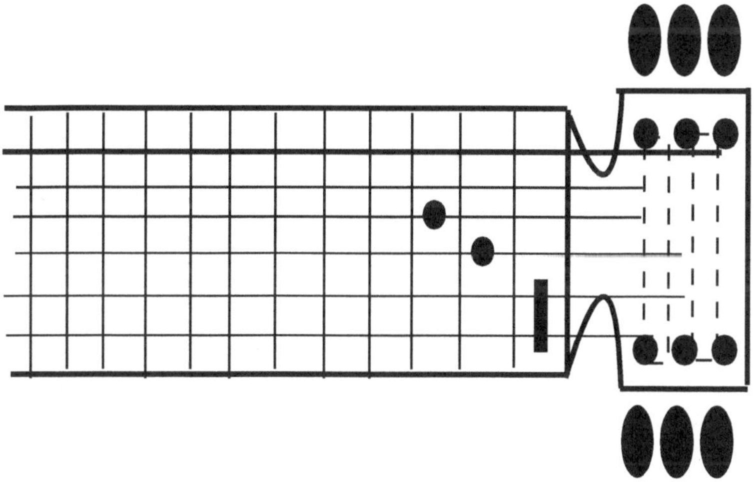

Guitar Notation Lab

The Nature of Technology

Technological Resourcses

Guitar Notation

Design Brief: Learn G major chord (First position).

G: ***E String, 3rd Fret***
B: ***A String, 2nd Fret***
D: ***D Open String***
G: ***G Open String***
B: ***B Open String***
G: ***E String 3rd Fret***

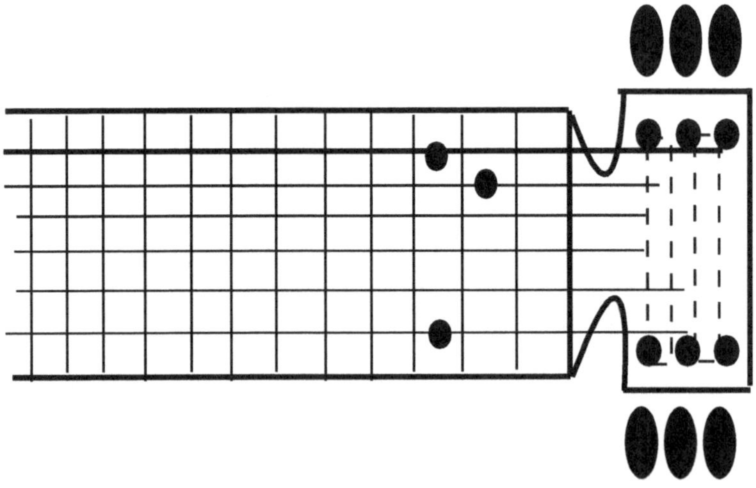

Guitar Notation Lab

The Nature of Technology

Technological Resourcses

Guitar Notation

Design Brief: Learn A minor chord (First position).

A: ***Open String***
E: **D *String, 2nd Fret***
D: ***G String, 2nd Fret***
G: ***G Open String** (Option 1st fret)*
C: ***B String, 1st Fret***

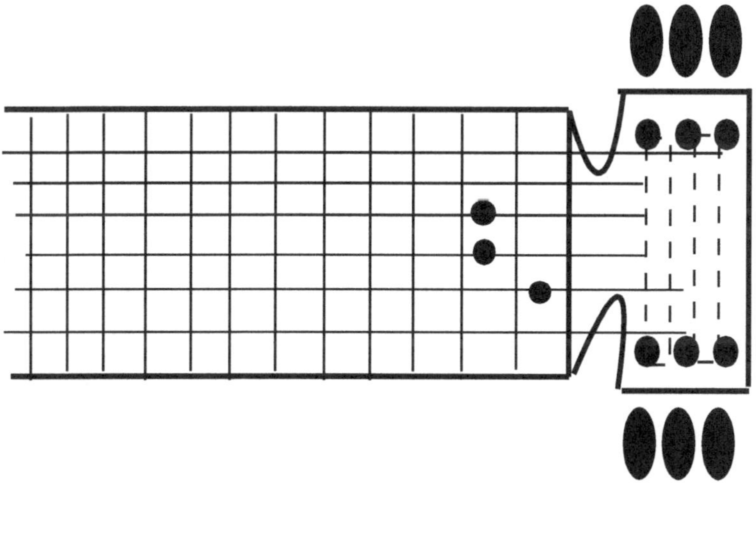

Guitar Notation Lab

The Nature of Technology

Technological Resourcses

Guitar Notation

Design Brief: Learn D minor chord (First position).

D: ***Open String***
A: ***G String, 2nd Fret***
D: ***B String, 3rd Fret***
F: ***E 1st Fret***

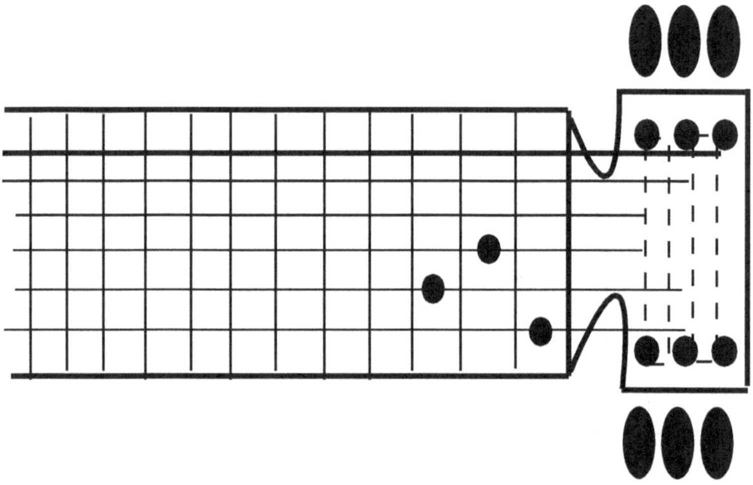

Guitar Notation Lab

The Nature of Technology

Technological Resourcses

Guitar Notation

Design Brief: Learn E minor chord (First position).

E6: ***Open String***
B: **A *String, 2nd Fret***
E: ***D String, 2nd Fret***
G: ***G Open String***
B: ***B Open String***
E: ***E Open String***

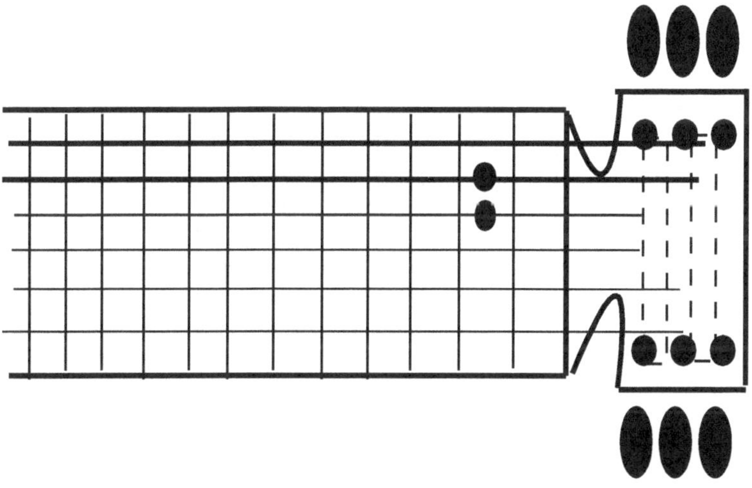

Guitar Notation Lab

The Nature of Technology

Technological Resourcses

Guitar Notation

Design Brief: Learn B minor 7-5 chord (First position).

B: ***A String, 2nd Fret***
D: ***D String, Open***
A: ***G String, 2nd Fret***
B: ***B String, 3rd Fret***
F: ***E String, 1st Fret***

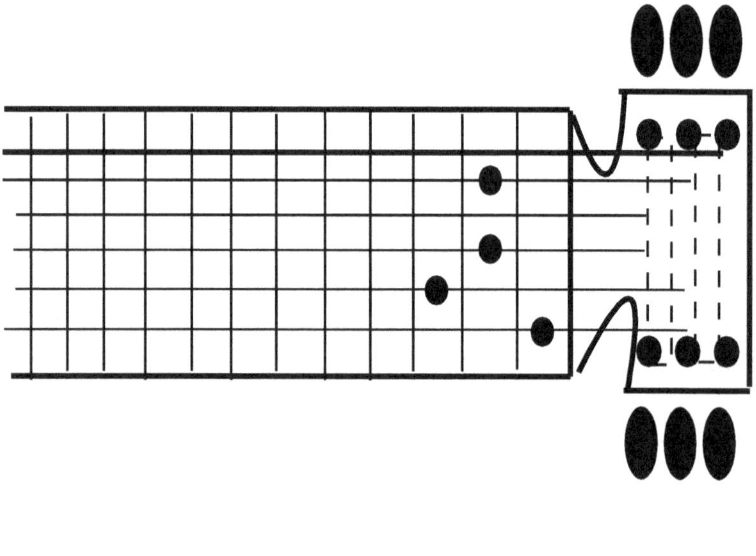

Songwriting Lab

Abilities in a Technological World

Possibility Statement: *STEM Educational Enrichment Services LLC needs a song promoting "STEM In Music".*

Design Brief: *Write a song in the key C major relating to "STEM In Music" Curriculum / Program.*

Live Your Life Today (Michael A. Johnson) © 2019

Keep your dreams alive,
Make plans to survive
Just know what it takes
to live your life.

Seek to find understanding,
Leading towards your destiny
Know what's right,
Know what's wrong,
Learn to choose and be strong

Put your best in what you Do.
Let it all come true for you.

Pick-up and go on your way,
Start living your life.

It's Time to know what to do.
It's Time to know what to say.
It's Time to know how to do.
Live Your Life Today.

it will carry you through time.
Carry you through time.
Make plans to survive.
it will carry you through time.

Carry you through time.

Come on and go your way,
find lessons on the way.
Keep Be-lie-ving what is right,
Never give up the fight,
today is just right.
Just keep on Lead-ing towards
a better day.

Better day,
Take steps towards a better day

Do you what you got to do.
Be what you want to be.
Live your life.

<u>Songwriting Lab</u>

Abilities in a Technological World

<u>Problem Statement:</u> *STEM Educational Enrichment Services LLC needs a song promoting "STEM In Music".*

<u>Design Brief:</u> *Write a song in the key C major relating to "STEM In Music" Curriculum / Program.*

<u>The Nature of life (Michael A. Johnson) © 2019</u>

Do you what you got to do.
(Live your Life today)
Be what you want to be.
(Let Your mind be free)
Never give up on you.
(Today is your life to live)
Let your hope survive.
It's <u>The Nature of Life.</u>

When you on and go your way,
Find lessons everyday,
Bringing you closer to your destiny.
Keep Be-lie-ving in you, and
Never give up hope, or dreams.
Today is just right for you to see.

Do you what you got to do.
(Live your Life today)
Be what you want to be.
(Let Your mind be free)
Never give up on you.
(Today is your life to live)
Let your hope survive.

It's The Nature of Life.

Do your Best in what you do.
Let your gifts lead the way.
Show the world better days.
Live with joy toady.
It's your moment in time.
Fight with a purpose.
Don't look back, be strong
Just keep moving on.

Do you what you got to do.
(Live your Life today)
Be what you want to be.
(Let Your mind be free)
Never give up on you.
(Today is your life to live)
Let your hope survive.
It's The Nature of Life.

Do you what you got to do.
(Live your Life today)
Be what you want to be.
(Let Your mind be free)
Never give up on you.
(Today is your life to live)
Let your hope survive.
It's The Nature of Life.

Songwriting Lab

Abilities in a Technological World

Problem Statement: *STEM Educational Enrichment Services LLC needs a song promoting "STEM In Music".*

Design Brief: *Write a song in the key C major relating to "STEM In Music" Curriculum / Program.*

Living in Society (Michael A. Johnson) © 2019

Be the best you can be,
(Living in Society).
Be the best you can be,
(Living in Society).

Keep on living. Know who you are.
Find your destiny in your heart.
Seek the truth, no matter how hard.
Never give up, be the best you can be.

Do you what you got to do.
Be what you want to be.
Live your life.
Living in society.
(Repeat):

Be the best you can be,
(Living in Society).
Be the best you can be,
(Living in Society).
(Repeat):

<u>Songwriting Lab</u>

Abilities in a Technological World

<u>**Problem Statement:**</u> *STEM Educational Enrichment Services LLC needs a song promoting "STEM In Music".*

<u>**Design Brief:**</u> *Write a song in the key C major relating to "STEM In Music" Curriculum / Program.*

<u>Celebrate Life (Michael A. Johnson) © 2019</u>

Share your love
With one another.
Give hope to each other.

<u>Celebrate Life</u>
to live right.
Come on let's celebrate today.

Come together for the
Good of man kind.
Come together
With our mind's.

Let's seek the good
in each other.
To seek the good
the good from
one another.

<u>Celebrate Life</u>
to live right.
Come on let's celebrate today.

Come together for the
Good of man kind.
Come together
With our mind's.
<u>(Repeat)</u>

<u>Songwriting Lab</u>

Abilities in a Technological World

<u>Problem Statement:</u> *STEM Educational Enrichment Services LLC needs a song promoting "STEM In Music".*

<u>Design Brief:</u> *Write a song in the key C major relating to "STEM In Music" Curriculum / Program.*

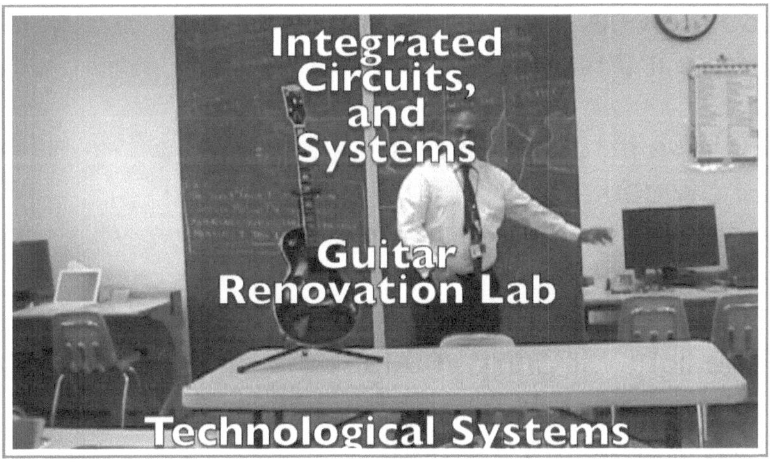

Rise Above The Moment (Michael A. Johnson) © 2019

Rise above the moment

This is the time.
Live with the greatness in you.

Stand-up, don't give-up.
This moment is made for you.

Live-up, don't give-up.
Be the one.
Make it through.

Believe in, don't give in
Show courage with strength.
This is the moment to Rise.

Reach Beyond the stars.
The light will carry you far.
When time gets hard.
Rise above the moment
(Rise Above this moment is for you).

Songwriting Lab

Abilities in a Technological World

Possibility Statement: *STEM Educational Enrichment Services LLC needs a song promoting "STEM In Music".*

Design Brief: *Write a song in the key C major relating to "STEM In Music" Curriculum / Program.*

Be Strong / Hold On (Michael A. Johnson) ©2019

No matter how hard it gets
You will find your place
In time.
Reach for a star above,
And fight to live your life

No matter how long it take
Each step brings you
Close to Destiny.
Your truth will set you free.

Don't Give Up Now.
You may feel alone
With No one to talk too.
Find the strength Deep inside.
It will carry you through Time
Be Strong, hold on.

Musical Instrument Invention Lab I

Technology and Society

Musical Instrument Invention

(Research & Development)

Safety

STEM In Music

(Musical Instrument Invention)

Before creating, making, or inventing anything, always remember, Safety first.

My Musical Instrument is specifically designed using safe manufacturing processes, materials, and electronic components. During Research and Developmental stages, safe proper use of tools, and equipment is paramount.

Lab Safety

Always follow instructions

Use tools, and equipment in a safe manner

Keep Lab area clean

Wear Safety Glasses, and Gloves during required Lab projects.

Always be careful working with wires (conductors / insulators), batteries, and integrated circuits & systems

Before connecting a circuit, or system make sure all power is turned off

Musical Instrument Invention Lab I

Technology and Society

Musical Instrument Invention
(Research & Development)
Invention & Innovation Design Process

Before learning, playing, or performing with a musical instrument; Invention / Innovation plays a major role. Science, Technology, Engineering, and Mathematics emulates vital role in designing, building, and manufacturing.

<u>Musical Instrument Invention Lab II</u>

Technology and Design

Musical Instrument Invention

(Research & Development)

<u>Invention & Innovation Design Process</u>

STEM In Music requirers designing a musical instrument using: Design Process, Possibility Statement, and Design Brief.

Design Process is a step by step procedure for collecting all information relating to design brief: Getting to Know The Possibilities, Exploring Concepts, Possible Solutions, Testing, and Evaluation.

Possibility Statement provides a specific need / want for producing a product, or device.

Design Brief defines possibility statement describing a solution towards achieving its goal.

<u>Musical Instrument Invention Lab II</u>

Technology and Design

Musical Instrument Invention

(Research & Development)

<u>Invention & Innovation Design Process</u>

<u>Problem Statement:</u>

STEM Educational Enrichment Services LLC "STEM In Music" needs a musical instrument invention.

<u>Design Brief:</u>

"STEM In Music" requires a new musical instrument using Principles of Engineering Concepts. This instrument will be used to write, produce, and record music. The learning process includes ITEEA (International Technology, Engineering for Educator Association), and TEEAP (Technology Engineering and Education Association of Pennsylvania) standards.

Getting to Know The Problem:

Create a sketch identifying a design for a new musical instrument.

Make a scale drawing (CAD) showing size, and measurements.

Implement ITEEA, and TEEAP Standards for Engineering Design process.

Test, evaluate, and present musical instrument invention.

Musical Instrument Invention Lab II

Technology and Design

Musical Instrument Invention

(Research & Development)

Invention & Innovation Design Process

List technological literacy requirements for musical instrument.

Identify technological system inventing musical instrument.

List Tools used for designing musical instruments.

Make a sketch showing size, shape, and measurements of new musical instruments.

Exploring Ideas:

Sketch three possible designs for a new music instrument.

Select best design, and additional modification if needed.

Identify Science, Technology, Engineering, Mathematics, and Musical concepts.

List resources, and materials needed for manufacturing.

Test / Evalute power, and energy requirements.

Musical Instrument Invention Lab II

Technology and Design

Musical Instrument Invention

(Research & Development)

Invention & Innovation Design Process

Possible Solutions:

Explain technological system operations for new musical instrument.

Accumulate all resources, and materials needed for manufacturing.

Select tools for designing musical instruments.

Create an electrical schematic sketch for musical instrument.

Testing & Evaluation

Use Pro Desktop (CAD Application) for final design.

Use Makerbot 3D Printer to manufacture music instrument.

Test musical instrument by playing.

Evaluate musical instrument.

Musical Instrument Invention

Technology and Design

Musical Instrument Invention

(Research & Development)

CAD (Computer Aided Design)

Computer Aided Design is the foundation developing a Musical Instrument Invention. The accuracy involving measurements, and technological systems help communicate design for manufacturing.

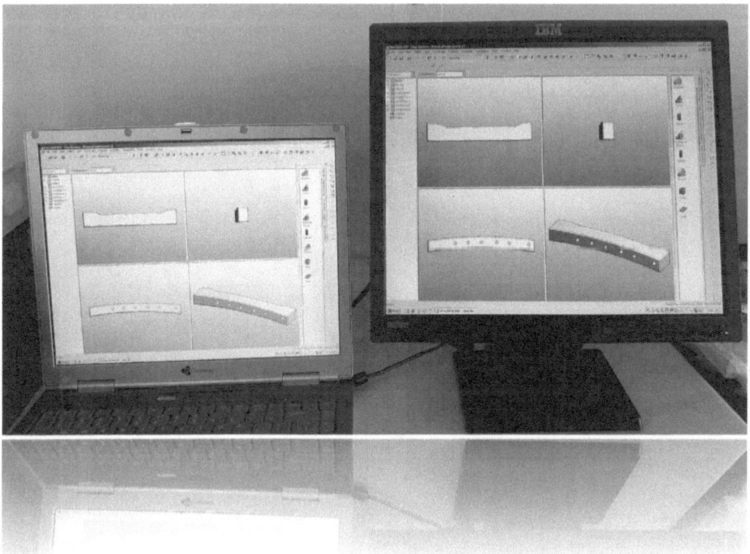

Musical Instrument Manufacturing

Technology and Design

Musical Instrument Invention

(Research & Development)

3D Modeling / Printing

3D Modeling Printer

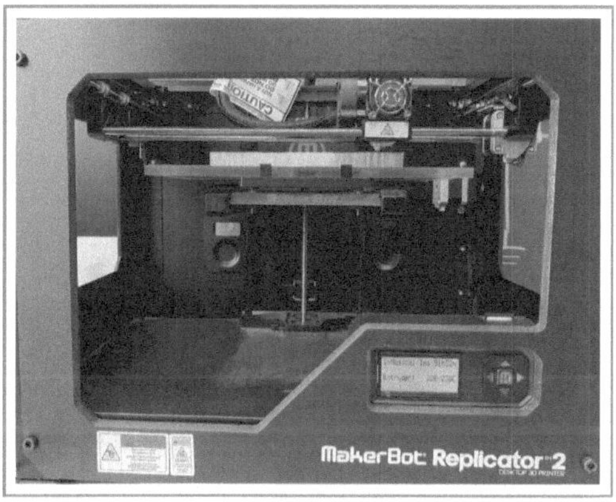

My Musical Instrument (m2i)

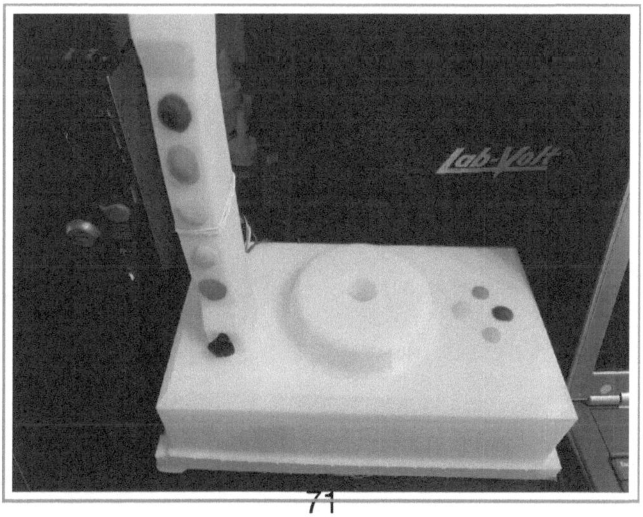

Musical Instrument Design

Design and Technology Education

Portfolio

Musical Instrument Invention
(Research & Development)
Problem Statement / Design Brief

Design and Technology Education

PROBLEM STATEMENT:

STEM EDUCATIONAL ENRICHMENT SERVICES LLC "STEM IN MUSIC" NEEDS A MUSICAL INSTRUMENT INVENTION.

DESIGN BRIEF:

"STEM IN MUSIC" REQUIRES A NEW MUSICAL INSTRUMENT USING PRINCIPLES OF ENGINEERING CONCEPTS. THIS INSTRUMENT WILL BE USED TO WRITE, PRODUCE, AND RECORD MUSIC. THE LEARNING PROCESS INCLUDES; ITEEA (INTERNATIONAL TECHNOLOGY AND ENGINEERING EDUCATOR ASSOCIATION), AND TEEAP (TECHNOLOGY ENGINEERING EDUCATION ASSOCIATION OF PENNSYLVANIA STANDARDS.

The Design Brief

1. **DEFINE** the problem:
   a) Describe what the designer (you) will do,
   and
   b) What the solution will accomplish
   and
   c) Identify any specifications that could help:
   - clarify
   - set limits
   - direct the design effort

Write/Draw your design brief (your own words & pictures) in the space provided.

GETTING TO KNOW THE PROBLEM:

- CREATE A SKETCH, PLAN, AND DESIGN OF A NEW MUSICAL INSTRUMENT.
- MAKE A SCALE DRAWING (CAD) SHOWING SIZE, AND MEASUREMENTS.
- IMPLEMENT ITEEA, AND TEEAP STANDARDS
- TEST, EVALUATE, AND PRESENT MUSICAL INSTRUMENT INVENTION.

NOTE: ADDITIONAL INFORMATION: AI-1

Design Portfolio

MICHAEL A. JOHNSON	MUSICAL INSTRUMENT INVENTION
	Project Title
	FALL SEMESTER 2019 STEM STUDIOS
Design Team Members	**Date** AUGUST 01, 2019 **Room #**

Musical Instrument Design

Design and Technology Education

Portfolio

Musical Instrument Invention
(Research & Development)
Explore the Problem / Possible Solution

EXPLORE THE PROBLEM

2. Investigation

To direct your research and help organize your efforts, you will need to ask and answer certain questions. Write your ????'s about the problem in the space provided.

Q1 SKETCH THREE POSSIBLE DESIGNS FOR NEW MUSICAL INSTRUMENT.

Q2 SELECT BEST DESIGN, AND ADDITIONAL MODIFICATION IF NEEDED.

Q3 IDENTIFY SCIENCE, TECHONOLOGY, ENGINEERING, MATHEMATI AND MUSICAL CONCEPTS.

Q4 LIST RESOURCES TOOLS, AND MATERIALS FOR MANUFACTURIN

Q5 POWER, AND ENERGY REQUIREMENTS.

---

3. Possible Solutions

In order to achieve the best possible solution, a number of ideas should be considered. Ideas can be far-fetched or very practical. Sometimes, a "crazy" idea leads to a very innovative answer.

Sketch & write ideas that come to mind as you brainstorm. As you do, you will begin to see variations or other ideas that will lead to better solutions. Let your first idea lead you to other ideas. Use the idea clouds to record your ideas as they come mind.

Make sure to list advantages and disadvantages of each idea. Remember, use pictures and words to explain each of your ideas.

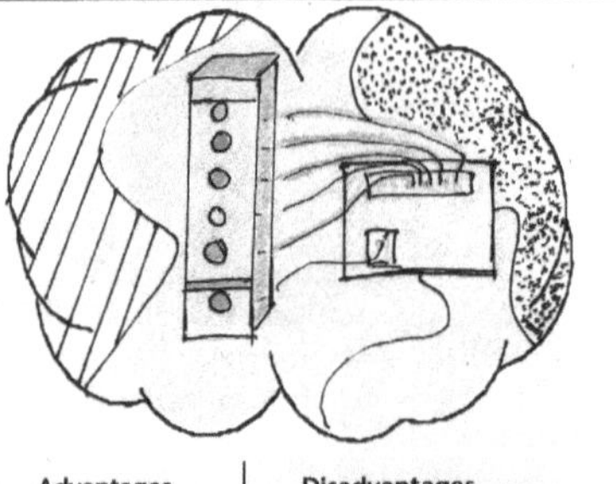

Advantages	Disadvantages
AESTHETICALLY DESIGNED.	LOCATION OF WIRES, AND INTEGRATED CIRCUIT.
EASY TO HOLD.	USB CONNECTIONS.
EXCELLENT FINGERING POSITION.	

Adv

EASY T

PLAY US

OR FING

SIZE, A

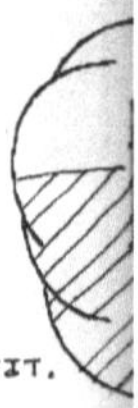

Musical Instrument Design

Design and Technology Education

Portfolio

Musical Instrument Invention

(Research & Development)

Pick & Plan / Possible Solution

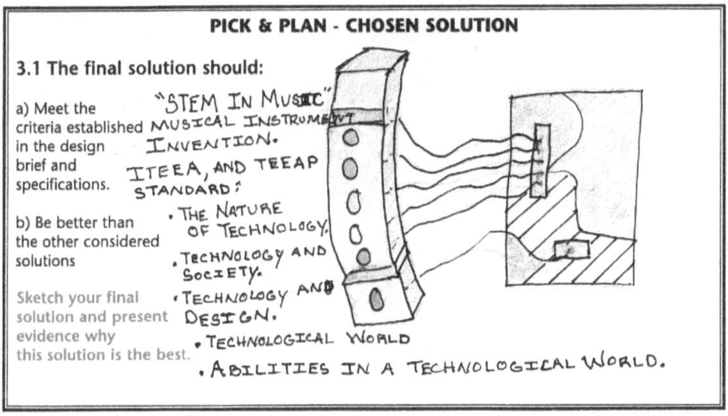

PICK & PLAN - CHOSEN SOLUTION

3.1 The final solution should:

a) Meet the criteria established in the design brief and specifications. "STEM IN MUSIC" MUSICAL INSTRUMENT INVENTION. ITEEA, AND TEEAP STANDARD:

b) Be better than the other considered solutions
- THE NATURE OF TECHNOLOGY.
- TECHNOLOGY AND SOCIETY.
- TECHNOLOGY AND DESIGN.
- TECHNOLOGICAL WORLD
- ABILITIES IN A TECHNOLOGICAL WORLD.

Sketch your final solution and present evidence why this solution is the best.

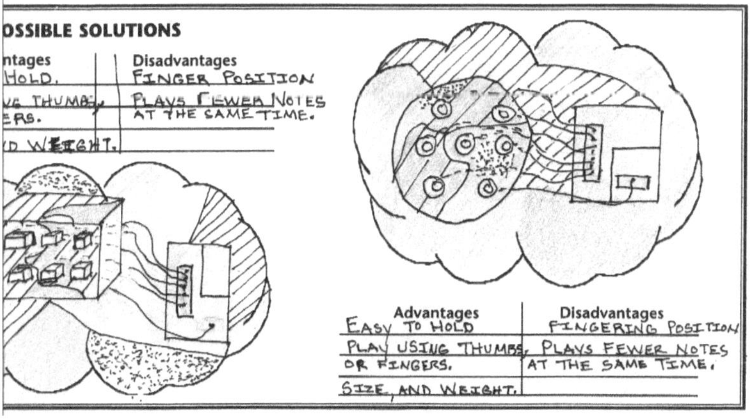

OSSIBLE SOLUTIONS

ntages	Disadvantages
HOLD.	FINGER POSITION
VG THUMBS,	PLAYS FEWER NOTES
ERS.	AT THE SAME TIME.
D WEIGHT.	

Advantages	Disadvantages
EASY TO HOLD	FINGERING POSITION
PLAY USING THUMBS,	PLAYS FEWER NOTES
OR FINGERS.	AT THE SAME TIME.
SIZE, AND WEIGHT.	

Musical Instrument Design

Design and Technology Education

Portfolio

Musical Instrument Invention

(Research & Development)

Plan & Produce I

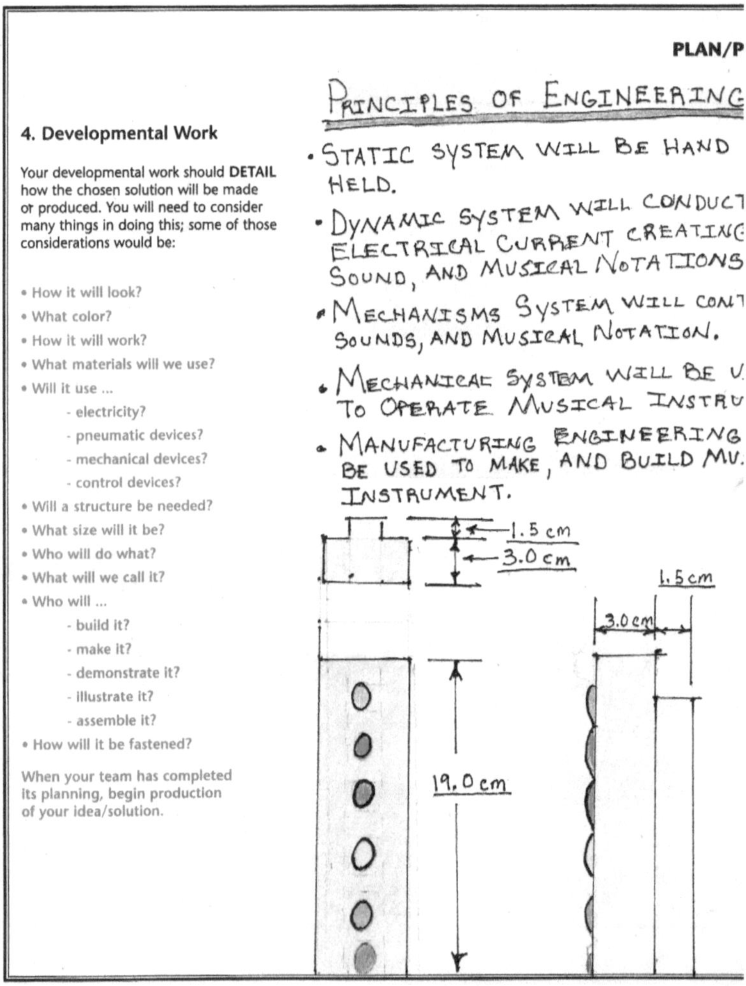

PLAN/P

4. Developmental Work

Your developmental work should **DETAIL** how the chosen solution will be made or produced. You will need to consider many things in doing this; some of those considerations would be:

- How it will look?
- What color?
- How it will work?
- What materials will we use?
- Will it use ...
  - electricity?
  - pneumatic devices?
  - mechanical devices?
  - control devices?
- Will a structure be needed?
- What size will it be?
- Who will do what?
- What will we call it?
- Who will ...
  - build it?
  - make it?
  - demonstrate it?
  - illustrate it?
  - assemble it?
- How will it be fastened?

When your team has completed its planning, begin production of your idea/solution.

PRINCIPLES OF ENGINEERING

- STATIC SYSTEM WILL BE HAND HELD.
- DYNAMIC SYSTEM WILL CONDUCT ELECTRICAL CURRENT CREATING SOUND, AND MUSICAL NOTATIONS
- MECHANISMS SYSTEM WILL CON[T] SOUNDS, AND MUSICAL NOTATION.
- MECHANICAL SYSTEM WILL BE U[SED] TO OPERATE MUSICAL INSTRU[MENT]
- MANUFACTURING ENGINEERING BE USED TO MAKE, AND BUILD MU[SICAL] INSTRUMENT.

Musical Instrument Design

Design and Technology Education

Portfolio

Power & Energy

Plan & Produce II

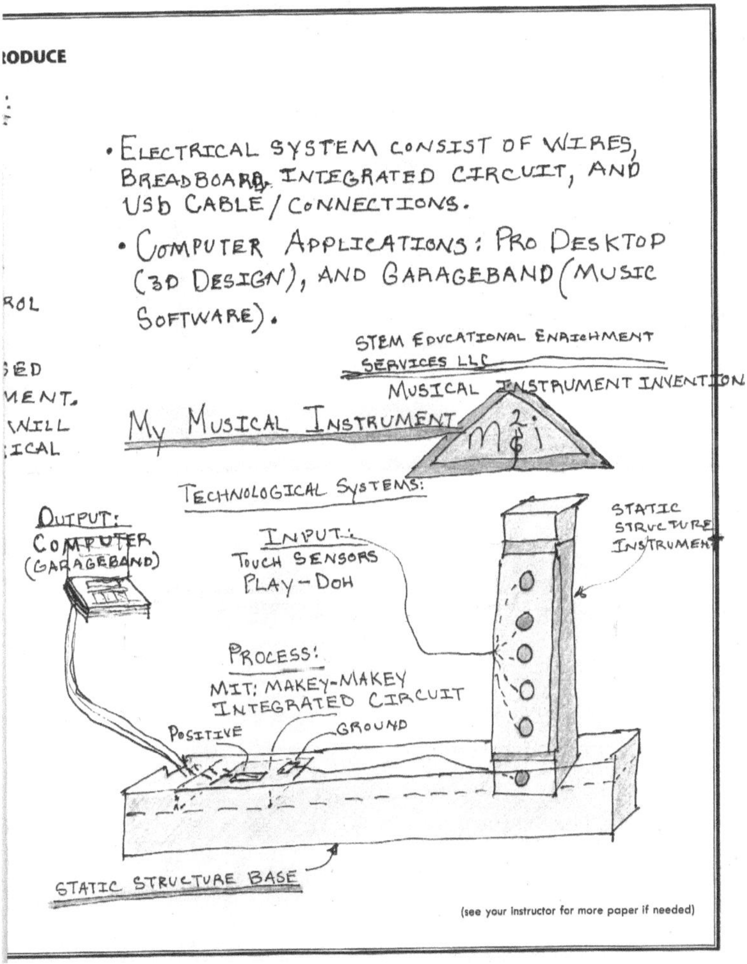

Musical Instrument Design

Design and Technology Education

Portfolio

Musical Instrument Invention

(Research & Development)

Power & Energy

Testing I

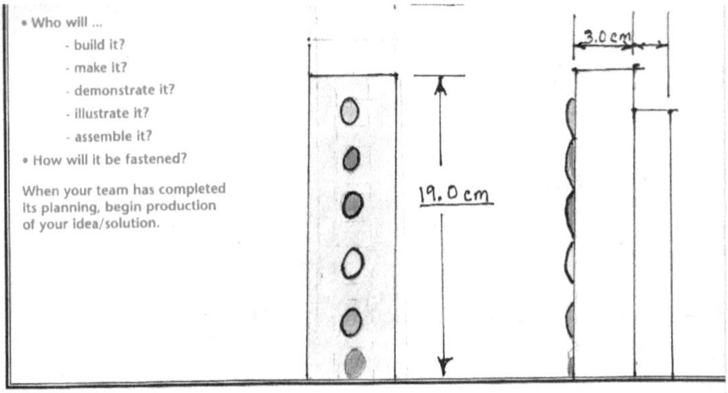

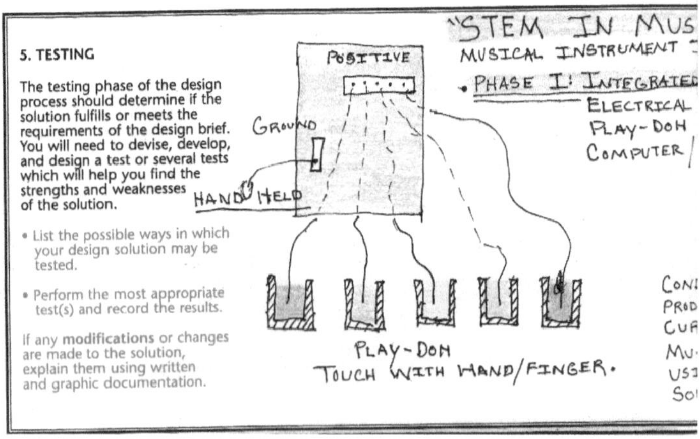

Musical Instrument Design

Design and Technology Education

Portfolio

(Research & Development)

Power & Energy

Testing II

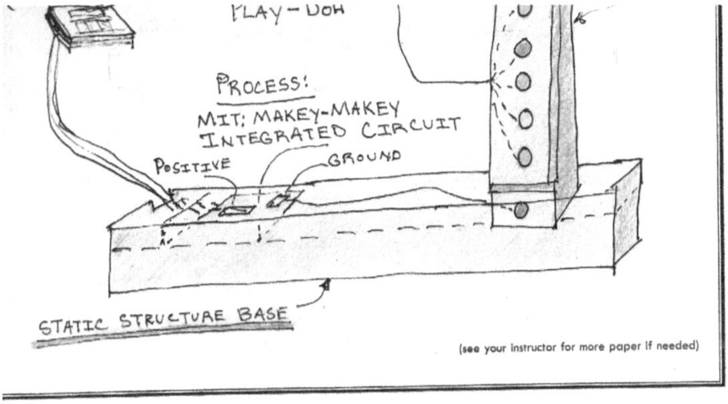

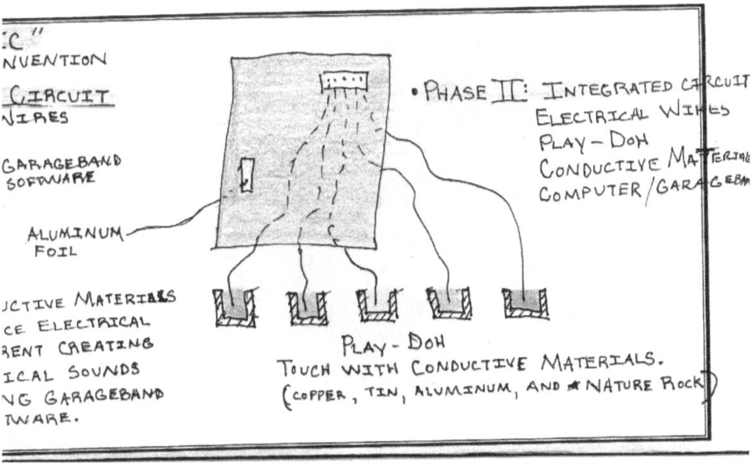

Musical Instrument Design

Design and Technology Education

Portfolio

Musical Instrument Invention
(Research & Development)
Power & Energy
Evaluations / Conclusions

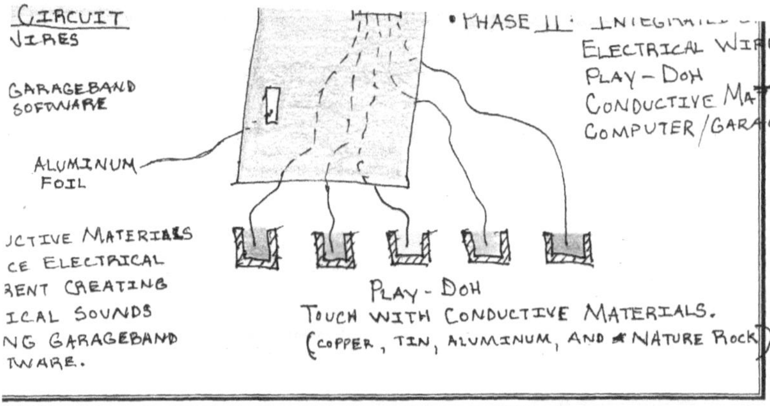

7. SELF-EVALUATION As a design team, ask yourselves the following questions and select the appropriate response. Pick one member of the team to lead the discussion on these questions and then to record the responses.

1. ~~Did we use our time effectively?~~ TIME EFFECTIVENESS 1 2 3 4 (5)
2. ~~Did we communicate well as a group?~~ ~~DESIGN~~ COMMUNICATION 2 3 (4) 5
3. ~~How well did we present our ideas?~~ DESIGN PRESENTATION 2 3 (4) 5
4. ~~Did the team reach its target?~~ DESIGN GOAL 1 2 3 4 (5)
5. What were the strengths and weaknesses of the final solution?

STRENGTHS: AESTHETICALLY DESIGN
FINGER POSITION
COMFORTABLE TO PLAY

WEAKNESS: CONNECTION INVOLVING STATIC, AND TECHNOLOGICAL SYSTEMS.

6. ~~As a team, what was the~~ hardest portion of this design activity?

THE INITIAL DESIGN STAGE, STATIC, AND DYNAMIC/TECHNOLOGICAL SYSTEMS WERE SEPERATE. THIS CAUSED PROBLEMS ~~WITH~~ CONNECTING ELECTRICAL WIRES.

Musical Instrument Design

Design and Technology Education

Portfolio

Musical Instrument Invention
(Research & Development)
Power & Energy
Evaluations / Conclusions

strengths and weaknesses of the solution.

- List the possible ways in which your design solution may be tested.
- Perform the most appropriate test(s) and record the results.

If any **modifications** or changes are made to the solution, explain them using written and graphic documentation.

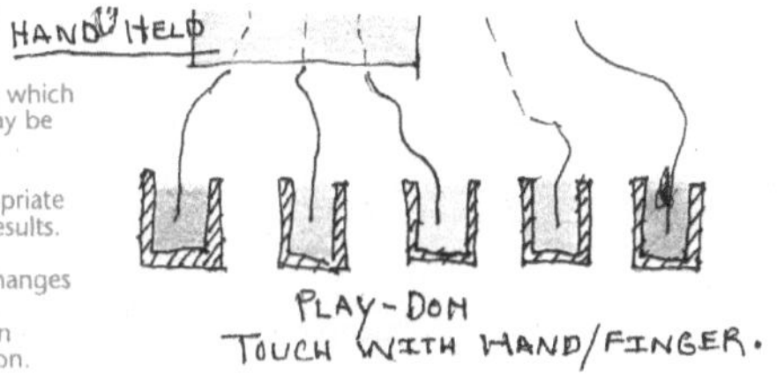

CON.
PROC
CUF
MU
US:
SO

EVALUATIONS/CONCLUSIONS

6.1. SUMMARIZE AND EVALUATE YOUR SOLUTION

Evaluate your design solution. Evaluation of your design solution will include how well your idea solved the design brief; therefore ...

- Did it carry the load?
- Did it go the distance?
- Did it work as designed?

STEM EDUCATIONAL ENRICHMENT SERVICES LLC
"STEM IN MUSIC"
MUSICAL INSTRUMENT INVENTION
PRESENTS:
"MY MUSICAL INSTRUMENT"

- STATIC SYSTEM SUPPORTS TOUCH SENSORS, ELECTRICAL WIRES, AND INTEGRATED CIRCUIT.
- TECHNOLOGICAL SYSTEM PLAYS MUSICAL NOTES.

6.2. FUTURE RESEARCH & DEVELOPMENT

Imagine...
You will make recommendations for further R & D, suggest improvements, or identify possible future uses.

STATIC SYSTEM, AND TECHNOLOGICAL SYSTEM DESIGNED AS ONE UNIT.

<u>*STEM Educational Enrichment Services LLC*</u>

"STEM In Music"

Principles of Engineering

Guitar Renovation

Learning Concepts

"STEM In Music" Principles of Engineering Learning Concepts include: Static, Dynamic, Load, Forces, Systems, and Mechanisms.

Guitar Renovation Static Structure:

Head: Guitar strings support system.

Neck: Guitar strings support system, and Frets.

Body: Guitar string support system, Electrical support system, and Mechanical support system.

Guitar Renovation Dynamic System

Head: Tuning Knobs, and Pegs

Neck: Guitar Strings

Body: Volume / Treble Controls, Toggle switch, and Bridge Mechanisms.

STEM Educational Enrichment Services LLC.

"STEM In Music"

Principles of Engineering

Guitar Renovation

Learning Concepts

Guitar Renovation Load, Effort, and Force

Head: Load, Effort, and Force of Tuning Knobs, and Pegs.

Neck: Load Effort, and Force of Guitar Strings.

Body: Load Effort, and Force of Volume / Treble Controls, Toggle switch, and Bridge Mechanisms.

<u>Guitar Renovation Force, Mass, and Gravitational / Applied Force</u>

Head: Force, Mass of Tuning Knobs, and Pegs.

Neck: Load Effort, and Force of Guitar Strings.

Body: Force & Mass (Weight) of
Volume / Treble Controls, Toggle switch,
and Bridge Mechanisms.

Total Guitar Force, and Mass (Weight).

STEM Educational Enrichment Services LLC

"STEM In Music"

Principles of Engineering

Guitar Renovation

Learning Concepts

Guitar Renovation Engineering Systems

Guitar Renovation Mechanisms

Head: Mechanical Mechanisms of Tuning Knobs, and Pegs.

Neck: Mechanical Mechanisms of Guitar Strings Support.

Body: Mechanical Mechanisms: Volume / Treble Controls, Toggle switch, and Bridge Mechanisms, an Pick-Ups.

STEM Educational Enrichment Services LLC.

"STEM In Music"

Principles of Engineering

Guitar Notation

Learning Concepts

"STEM In Music" Principles of Engineering Learning Concepts include: Static, Dynamic, Load, Forces, Systems, and Mechanisms.

Guitar Notation Static Structure Concepts:

Head: Tuning Guitar.

Neck: Note positions on Frets.

Body: Holding guitar in the best comfortable position.

Guitar Notation Dynamic System Concepts:

Head: Using Tuning Knobs to tune guitar

Neck: Positioning Fingers to play a note.

Body: Playing notes using finger, thumb, or pick.

STEM Educational Enrichment Services LLC.

"STEM In Music"

Principles of Engineering

Guitar Renovation Notation

Learning Concepts

Guitar Renovation / Notation Load, Effort, and Force

Head: Load, Effort, and Force of Tuning each String.

Neck: Load Effort, and Force Playing Guitar Strings.

Body: Load Effort, and Force adjusting Volume / Treble Controls.

Guitar Renovation / Notation Force, Mass, and Gravitational /Applied Force

Head: Force, and Mass of Guitar Strings.

Neck: Force, and Mass of Guitar Neck.

Body: Force, and Mass of Guitar body.

STEM Educational Enrichment Services LLC.

"STEM In Music"

Principles of Engineering

Guitar Notation

Learning Concepts

Guitar Notation Systems

Head: Adjusting Tuning Knobs for accurate tuning.

Neck: Technics playing notes, and chords.

Body: Adjusting Volume / Treble Controls and Toggle switch creating variation of tones.

Guitar Notation Mechanisms

Head: Adjusting Tuning Knobs, and Pegs for Tuning.

Neck: Adjusting fingering positioning to Play notes, and chords.

Body: Adjusting Mechanical Mechanisms for Volume / Treble Controls, Toggle switch, and Bridge Mechanisms.

<u>*STEM Educational Enrichment Services LLC.*</u>

"STEM In Music"

Principles of Engineering

Musical Instrument Invention

Learning Concepts

"STEM In Music" Principles of Engineering Musical Instrument Invention include: Static, Dynamic, Load, Forces, Systems, and Mechanisms.

Musical Instrument Invention Static Concepts:

Handheld Static Structure.

Electrical Structure Support.

Mechanical Static Structure.

Musical Instrument Invention Dynamic Concepts:

Handheld fingering position

Electrical Current / Flow

<u>STEM Educational Enrichment Services LLC.</u>

"STEM In Music"

Principles of Engineering

Musical Instrument Invention

Learning Concepts

Musical Instrument Invention Loads, Effort, Fulcrum, and Forces:

Mechanical Concept: Force, and Mass of Musical Instrument Invention.

Electrical Concept: Voltage,Current, and Resistance (Ohms Law) of Musical Instrument Invention.

Structural Concepts: Load, Effort, and Fulcrum of Musical Instrument Invention.

Musical Instrument Invention Systems:

Technological System: Input, Process, Output, and Feedback of Musical Instrument Invention.

Electrical System: Integrated Circuit, and System of Musical Instrument Invention.

Structural System: Sketch, and CAD (Computer Aided Design) Design of Musical Instrument Invention.

Musical Instrument Invention Mechanisms:

Mechanical Mechanisms: Volume / Tone controls of Musical Instrument Invention.

Electrical System: Integrated Circuits, and Systems of Musical Instrument Invention.

Structural System: 3D Modeling / Printing.

STEM Educational Enrichment Services LLC.

"STEM In Music"

Principles of Engineering

Musical Instrument Invention

Learning Concepts

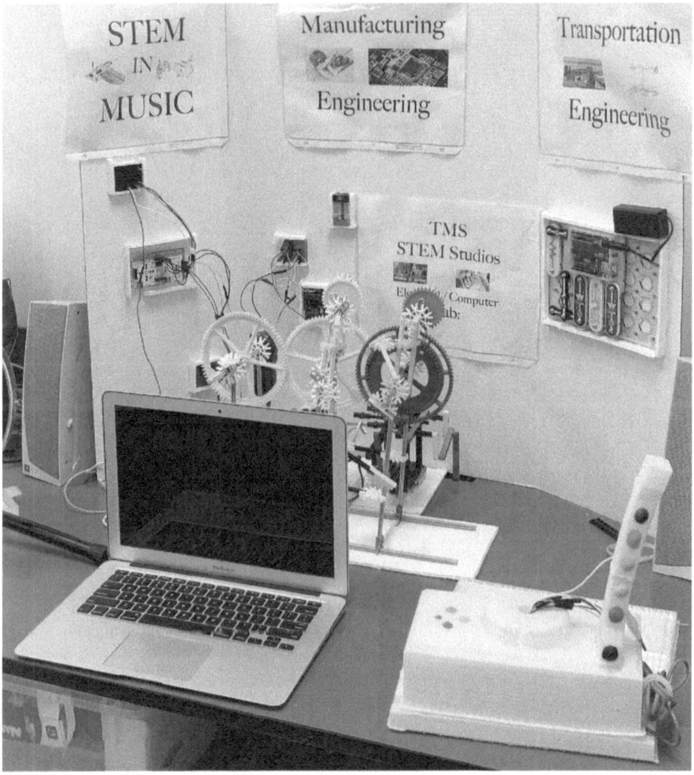

STEM Educational Enrichment Services LLC.

"STEM In Music"

Principles of Engineering

STEM Lab Lessons & Activities

STEM Lab Lessons:

1. *Static*
2. *Dynamic*
3. *Loads*
4. *Forces*
5. *Mechanisms*

STEM Lab Activities:

1. *Lab Briefing*
2. *Pre-Survey*
3. *Design Lab*
4. *Material Lab*
5. *Structural Lab*
6. *Mechanical Lab*
7. *Mid-Term Review*
8. *Electrical Engineering Lab*
9. *Manufacturing Lab (3D Modeling)*
10. *Mechanical Advantage*
11. *Lab Debriefing*
12. *Math Lab*
13. *Post Survey*

Discovery Design Brief:
1.01-06-24 / 745A
STEM In Education
Principles of Engineering
Teaching, and Learning Concepts
"STEM In Music"
Key of C Major *(Notes, and Chords)*

I: Possibility Statement:

Exploring the key of C Major increases understanding in music theory concepts, and applications.

II: Getting to Know The Possibilities:

1. C Major Diatonic Scale
2. C Major Pentatonic Scale
3. C Major Arpeggio Scale
4. C Major Pattern / Structure Progressions

III: Exploring Concepts:

1. C Major Diatonic Scale: 7 Notes, No Flats, or Sharp.
   C D E F G A B
2. C Major Pentatonic Scale: 5 Notes, No Flats, or Sharp.
   C D E G A
3. C Major Arpeggio Scale: 3 Notes, No Flats, or Sharp.
   C E G
4. C Major Diatonic Chords:
   C^M D^m E^m F^M G^M A^m B°
5. C Major (Major Chords):
   C F G
6. C Major (Minor Chords):
   A D E

7. C Major (Diminish Chord):
   B°

IV: Possibile Solutions:

C Major "Major" Chord Progressions

1. C^M	F^M	G^M
2. C^M	G^M	F^M
3. F^M	C^M	G^M
4. G^M	C^M	F^M
5. F^M	G^M	C^M
6. G^M	F^M	C^M

C Major "Minor" Chord Progressions

1. Am	Dm	Em
2. Am	Em	Dm
3. Dm	Am	Em
4. Em	Am	Dm
5. Dm	Em	Am
6. Em	Dm	Am

C Major "Diminish Chord Progressions

1. CM	B°	CM
2. FM	B°	CM
3. GM	B°	CM
4. Cm	B°	Am
5. Am	B°	CM

V: Testing, and Evaluation:

C Major Chord Progressions

1. C^M	Am	Dm	G^M
2. Am	Dm	G^M	C^M
3. F^M	Am	Dm	G^M
4. Am	Dm	G^M	F^M
5. Am	Dm	F^M	G^M
6. Dm	G^M	F^M	Am
7. C^M	Em	B°	Am
8. F^M	G^M	B°	C^M

Discovery Design Brief:
2.01-06-24 / 1047A
STEM In Education
Principles of Engineering
Teaching, and Learning Concepts
"STEM In Music"
Key of A minor (Notes, and Chords)

I: Possibility Statement:

Exploring the key of A minor increases understanding in music theory concepts, and applications.

II: Getting to Know The Possibilities:

1. A minor Diatonic Scale
2. A minor Pentatonic Scale
3. A minor Arpeggio Scale
4. A minor Pattern / Structure Progressions

III: Exploring Concepts:

1. A minor Diatonic Scale: 7 Notes, No Flats, or Sharp.
   A B C D E F
2. A minor Pentatonic Scale: 5 Notes, No Flats, or Sharp.
   A C D E G
3. A minor Arpeggio Scale: 3 Notes, No Flats, or Sharp.
   A C E
4. A minor Diatonic Chords:
   Am B° CM Dm Em FM GM
5. A minor (Major Chords):
   C F G
6. A minor (Minor Chords):
   A D E
7. A minor (Diminish Chord):
   B°

IV: Possibile Solutions:

A minor "Minor" Chord Progressions

1. Am	Dm	Em
2. Am	Em	Dm
3. Dm	Am	Em
4. Em	Am	Dm
5. Em	Dm	Am
6. Dm	Em	Am

A minor "Major" Chord Progressions

1. C^M	F^M	G^M
2. C^M	G^M	F^M
3. F^M	C^M	G^M
4. G^M	C^M	F^M
5. F^M	G^M	C^M
6. G^M	F^M	C^M

A minor "Diminish Chord Progressions

1. Am	B°	Am
2. Dm	B°	Am
3. Em	B°	Am
4. C^M	B°	C^M
5. C^M	B°	Am

V: Testing, and Evaluation:

A minor / C Major Chord Progressions

1. Am	C^M	F^M	G^M
2. Am	F^M	C^M	G^M
3. Dm	C^M	G^M	Am
4. C^M	F^M	G^M	Dm
5. C^M	F^M	Dm	G^M
6. F^M	G^M	Dm	C^M
7. Am	G^M	B°	C^M
8. Dm	G^M	B°	Am

Discovery Design Brief:
3.01-07-24 / 342A
STEM In Education
Principles of Engineering
Teaching, and Learning Concepts
"STEM In Music"
Key of F Major (Notes, and Chords)

I: Possibility Statement:

Exploring the key of F Major increases understanding in music theory concepts, and applications.

II: Getting to Know The Possibilities:

1. F Major Diatonic Scale
2. F Major Pentatonic Scale
3. F Major Arpeggio Scale
4. F Major Pattern / Structure Progressions

III: Exploring Concepts:

1. F Major Diatonic Scale: 7 Notes, 1 Flats.
   F G A Bb C D E
2. F Major Pentatonic Scale: 5 Notes, No Flats.
   F G A C D
3. F Major Arpeggio Scale: 3 Notes, No Flat.
   F A C
4. F Major Diatonic Chords:
   F^M G^m A^m B^M C^M D^m E°
5. F Major (Major Chords):
   F^M Bb^M C^M
6. F Major (Minor Chords):
   Gm Am Dm
7. F Major (Diminish Chord):
   E°

IV: Possibile Solutions:

F Major "Major" Chord Progressions

1. F^M	Bb^M	C^M
2. Bb^M	C^M	F^M
3. C^M	F^M	Bb^M
4. C^M	F^M	Bb^M
5. Bb^M	C^M	F^M
6. C^M	Bb^M	F^M

F Major "Minor" Chord Progressions

1. Dm	Am	Gm
2. Dm	Gm	Am
3. Dm	Am	Dm
4. Gm	Am	Dm
5. Gm	Dm	Am
6. Am	Gm	Dm

F Major "Diminish Chord Progressions

1. F^M	E°	F^M
2. Bb^M	E°	F^M
3. F^M	E°	Bb^M
4. Fm	E°	Dm
5. Dm	E°	F^M

V: Testing, and Evaluation:

F Major / D minor Chord Progressions

1. FM	Dm	Bb^M	C^M
2. Dm	Bb^M	FM	C^M
3. Bb^M	F^M	C^M	Dm
4. F^M	Bb^M	C^M	Dm
5. F^M	Bb^M	Gm	C^M
6. Bb^M	CM	Gm	F^M
7. Bb^M	C^M	E°	F^M
8. Dm	G^M	E°	F^M

Discovery Design Brief:
4.01-07-24 / 745A
STEM In Education
Principles of Engineering
Teaching, and Learning Concepts
"STEM In Music"
Key of D minor (Notes, and Chords)

I: Possibility Statement:

Exploring the key of D minor increases understanding in music theory concepts, and applications.

II: Getting to Know The Possibilities:

1. D minor Diatonic Scale
2. D minor Pentatonic Scale
3. D minor Arpeggio Scale
4. D minor Pattern / Structure Progressions

III: Exploring Concepts:

1. D minor Diatonic Scale: 7 Notes, 1 Flat.
   D E F G A Bb C
2. D minor Pentatonic Scale: 5 Notes, No Flats.
   D F G A C
3. D minor Arpeggio Scale: 3 Notes, No Flats.
   D F A
4. D minor Diatonic Chords:
   Dm E° FM Gm Am B^bM C^M
5. D minor (Major Chords):
   F Bb C
6. D minor (Minor Chords):
   D G A
7. D minor (Diminish Chord):
   E°

IV: Possibile Solutions:

D minor "Minor" Chord Progressions

1. Dm	Gm	Am
2. Dm	Am	Gm
3. Gm	Gm	Dm
4. Gm	Dm	Am
5. Am	Dm	Gm
6. Am	Gm	Dm

D minor "Major" Chord Progressions

1. F^M	Bb^M	C^M
2. F^M	C^M	Bb^M
3. Bb^M	F^M	C^M
4. C^M	Bb^M	F^M
5. C^M	F^M	Bb^M
6. Bb^M	C^M	F^M

D minor "Diminish Chord Progressions

1. Dm	E^o	Dm
2. Gm	E^o	Gm
3. Am	E^o	Am
4. G^M	E^o	G^M
5. Bb^M	E^o	G^M

V: Testing, and Evaluation:

D minor / F Major Chord Progressions

1. Dm	F^M	Bb^M	C^M
2. Dm	Bb^M	C^M	F^M
3. F^M	C^M	Bb^M	Am
4. Bb^M	F^M	C^M	Gm
5. C^M	F^M	Dm	Bb^M
6. F^M	C^M	Gm	Bb^M
7. Am	Bb^M	E°	F^M
8. Gm	C^M	E°	Dm

Discovery Design Brief:
5.01-07-24 / 834A
STEM In Education
Principles of Engineering
Teaching, and Learning Concepts
"STEM In Music"
Key of G Major (Notes, and Chords)

I: Possibility Statement:

Exploring the key of G Major increases understanding in music theory concepts, and applications.

II: Getting to Know The Possibilities:

1. G Major Diatonic Scale
2. G Major Pentatonic Scale
3. G Major Arpeggio Scale
4. G Major Pattern / Structure Progressions

III: Exploring Concepts:

1. G Major Diatonic Scale: 7 Notes, 1 Sharp.
   G A B C D E F#
2. G Major Pentatonic Scale: 5 Notes, No Sharps.
   G A B D E
3. G Major Arpeggio Scale: 3 Notes, No Sharps.
   G B D
4. G Major Diatonic Chords:
   G^M A^m B^m C^M D^M E^m F#°
5. G Major (Major Chords):
   G^M C^M D^M
6. G Major (Minor Chords):
   Am Bm Em
7. G Major (Diminish Chord):
   F#°

IV: Possibile Solutions:

G Major "Major" Chord Progressions

1. G^M	C^M	D^M
2. G^M	D^M	C^M
3. C^M	G^M	D^M
4. D^M	G^M	C^M
5. C^M	D^M	G^M
6. D^M	C^M	G^M

G Major "Minor" Chord Progressions

1. Em	Am	Bm
2. Em	Bm	Am
3. Am	Bm	Em
4. Am	Em	Bm
5. Bm	Em	Am
6. Bm	Am	Em

G Major "Diminish Chord Progressions

1. G^M	F#°	G^M
2. C^M	F#°	G^M
3. G^M	F#°	C^M
4. G^M	F#°	D^M
5. D^M	F#°	G^M

V: Testing, and Evaluation:

G Major / E minor Chord Progressions

1. G^M	Em	C^M	D^M
2. Em	C^M	G^M	D^M
3. C^M	GM	D^M	Em
4. G^M	C^M	D^M	Em
5. G^M	C^M	Am	D^M
6. C^M	D^M	Bm	G^M
7. C^M	CD^M	F#°	G^M
8. Em	G^M	F#°	G^M

<u>Discovery Design Brief:</u>
<u>6.01-19-24 / 1221P</u>
STEM In Education
Principles of Engineering
Teaching, and Learning Concepts
<u>"STEM In Music"</u>
<u>Key of E minor</u> (Notes, and Chords)

<u>I: Possibility Statement:</u>

Exploring the key of E minor increases understanding in music theory concepts, and applications.

<u>II: Getting to Know The Possibilities:</u>

1. E minor Diatonic Scale
2. E minor Pentatonic Scale
3. E minor Arpeggio Scale
4. E minor Pattern / Structure Progressions

<u>III: Exploring Concepts:</u>

1. E minor Diatonic Scale: 7 Notes, 1 Sharp.
   E F# G A B C D
2. E minor Pentatonic Scale: 5 Notes, No Sharps.
   E G A B C
3. E minor Arpeggio Scale: 3 Notes, No Sharps.
   E G B
4. E minor Diatonic Chords:
   Em F#° GM Am Bm C^M D^M
5. E minor (Major Chords):
   D B^b C
6. E minor (Minor Chords):
   E A B
7. E minor (Diminish Chord):
   F#°

<u>IV: Possibile Solutions:</u>

E minor "Minor" Chord Progressions

1. Em	Am	Bm
2. Em	Bm	Am
3. Am	Bm	Em
4. Bm	Em	Am
5. Am	Em	Bm
6. Bm	Am	Em

E minor "Major" Chord Progressions

1. G^M	C^M	D^M
2. G^M	D^M	C^M
3. C^M	D^M	G^M
4. C^M	G^M	D^M
5. D^M	G^M	C^M
6. D^M	C^M	G^M

E minor "Diminish Chord Progressions

1. Em	F#°	Em
2. Am	F#°	Am
3. Bm	F#°	Bm
4. G^M	F#°	Em
5. Em	F#°	G^M

V: Testing, and Evaluation:

E minor / G Major Chord Progressions

1. Em	G^M	C^M	D^M
2. Em	C^M	G^M	D^M
3. Am	G^M	D^M	Em
4. G^M	C^M	D^M	Am
5. G^M	C^M	Am	D^M
6. C^M	D^M	Am	D^M
7. Gm	G^M	F#°	G^M
8. Bm	D^M	F#°	Em

Discovery Design Brief:
7.01-31-24 / AP
STEM In Education
Principles of Engineering
Teaching, and Learning Concepts
"STEM In Music"
Key of Bb Major (Notes, and Chords)

I: Possibility Statement:

Exploring the key of Bb Major increases understanding in music theory concepts, and applications.

II: Getting to Know The Possibilities:

1. Bb Major Diatonic Scale
2. Bb Major Pentatonic Scale
3. Bb Major Arpeggio Scale
4. Bb Major Pattern / Structure Progressions

III: Exploring Concepts:

1. Bb Major Diatonic Scale: 2 Flats.
   Bb C D Eb F G G
2. Bb Major Pentatonic Scale: 5 Notes, 1 Flat.
   Bb C D F G
3. Bb Major Arpeggio Scale: 3 Notes, 1 Flat.
   Bb D F
4. Bb Major Diatonic Chords:
   Bb^M C^m D^m Eb^M F^M G^m A°
5. Bb Major (Major Chords):
   Bb^M Eb^M F^M
6. Bb Major (Minor Chords):
   Cm Dm Gm

7. Bb Major (Diminish Chord):
   A°

IV: Possibile Solutions:

Bb Major "Major" Chord Progressions

1. Bb^M	Eb^M	F^M
2. Eb^M	Bb^M	F^M
3. Bb^M	F^M	Eb^M
4. Eb^M	Bb^M	F^M
5. Eb^M	F^M	Bb^M
6. F^M	Eb^M	Bb^M

Bb Major "Minor" Chord Progressions

1. Gm	Cm	Dm
2. Cm	Dm	Gm
3. Dm	Cm	Gm
4. Dm	Gm	Cm
5. Gm	Dm	Cm
6. Cm	Gm	Dm

Bb Major "Diminish Chord Progressions

1. Bb^M	A°	Bb^M
2. Eb^M	A°	Eb^M
3. F^M	A°	Eb^M
4. Eb^M	A°	Gm
5. Gm	A°	Bb^M

V: Testing, and Evaluation:

Bb Major / G minor Chord Progressions

1. Bb^M	Gm	Eb^M	F^M
2. Gm	Bb^M	F^M	Eb^M
3. Eb^M	F^M	Bb^M	Gm
4. F^M	Bb^M	Eb^M	Gm
5. Eb^M	Bb^M	Gm	F^M
6. Bb^M	Eb^M	Gm	F^M
7. Eb^M	Bb^M	A°	Gm
8. Gm	Eb^M	A°	Bb^M

Discovery Design Brief:
8.01-31-24 / AP
STEM In Education
Principles of Engineering
Teaching, and Learning Concepts
"STEM In Music"
Key of G minor (Notes, and Chords)

I: Possibility Statement:
Exploring the key of G minor increases understanding in music theory concepts, and applications.

II: Getting to Know The Possibilities:
1. G minor Diatonic Scale
2. G minor Pentatonic Scale
3. G minor Arpeggio Scale
4. G minor Pattern / Structure Progressions

III: Exploring Concepts:
1. G minor Diatonic Scale: 2 Sharps.
   G A Bb C D Eb F
2. G minor Pentatonic Scale: 1 Sharp.
   G Bb C D F
3. G minor Arpeggio Scale: 3 Notes, 1 Sharp.
   G Bb D
4. G minor Diatonic Chords:
   Gm A° BbM Cm Dm Eb^M F^M
5. G minor (Major Chords):
   Bb Eb F
6. G minor (Minor Chords):
   G C D
7. G minor (Diminish Chord):
   A°

IV: Possibile Solutions:

G minor "Minor" Chord Progressions

1. Gm	Cm	Dm
2. Gm	Dm	Cm
3. Dm	Cm	Gm
4. Dm	Gm	Cm
5. Cm	Dm	Gm
6. Cm	Gm	Dm

G minor "Major" Chord Progressions

1. Bb^M	Eb^M	F^M
2. Bb^M	F^M	Eb^M
3. Eb^M	F^M	Bb^M
4. Eb^M	Bb^M	F^M
5. F^M	Bb^M	Eb^M
6. F^M	Eb^M	Bb^M

G minor "Diminish Chord Progressions

1. Gm	A°	Gm
2. Cm	A°	Bm
3. Dm	A°	Em
4. F#^M	A°	Bm
5. Bm	A°	D^M

V: Testing, and Evaluation:

G minor / Bb Major Chord Progressions

1. Gm	Bb^M	F^M	Eb^M
2. Cm	F^M	Eb^M	Bb^M
3. Dm	Eb^M	Bb^M	F^M
4. F^M	Bb^M	F^M	Gm
5. Eb^M	F^M	Cm	Bb^M
6. Bb^M	Eb^M	Gm	F^M
7. Gm	G^M	C#°	Bb^M
8. Dm	Bb^M	C#°	Gm

Discovery Design Brief:
9.01-29-24 / 257P
STEM In Education
Principles of Engineering
Teaching, and Learning Concepts
"STEM In Music"
Key of D Major (Notes, and Chords)

I: Possibility Statement:
Exploring the key of D Major increases understanding in music theory concepts, and applications.

II: Getting to Know The Possibilities:
1. D Major Diatonic Scale
2. D Major Pentatonic Scale
3. D Major Arpeggio Scale
4. D Major Pattern / Structure Progressions

III: Exploring Concepts:
1. D Major Diatonic Scale: 7 Notes, 2 Sharps.
   D E F# G A B C#
2. D Major Pentatonic Scale: 1 Sharp.
   D E F# A B
3. D Major Arpeggio Scale: 3 Notes, 1 Sharp.
   D F# A
4. D Major Diatonic Chords:
   D^M E^m F#^m G^M A^M B^m C#°
5. D Major (Major Chords):
   D^M G^M A^M
6. D Major (Minor Chords):
   Em F#m Bm

7. D Major (Diminish Chord):
   C#°

IV: Possibile Solutions:

D Major "Major" Chord Progressions

1. D^M	G^M	A^M
2. G^M	A^M	D^M
3. A^M	D^M	G^M
4. A^M	G^M	D^M
5. G^M	D^M	A^M
6. D^M	A^M	G^M

D Major "Minor" Chord Progressions

1. Bm	Em	F#m
2. Em	F#m	Bm
3. F#m	Bm	Em
4. F#m	Em	Bm
5. Bm	F#m	Em
6. Em	Bm	F#m

D Major "Diminish Chord Progressions

1. D^M	C#°	D^M
2. G^M	C#°	D^M
3. A^M	C#°	D^M
4. G^M	C#°	A^M
5. A^M	C#°	G^M

V: Testing, and Evaluation:

D Major / B minor Chord Progressions

1. D^M	Bm	A^M	G^M
2. Bm	G^M	A^M	D^M
3. G^M	A^M	D^M	Em
4. G^M	A^M	D^M	F#m
5. A^M	G^M	F#m	D^M
6. G^M	D^M	Bm	A^M
7. A^M	D^M	C#°	D^M
8. Em	A^M	C#°	G^M

Discovery Design Brief:
10.01-31-24 / AP
STEM In Education
Principles of Engineering
Teaching, and Learning Concepts
"STEM In Music"
Key of B minor (Notes, and Chords)

I: Possibility Statement:

Exploring the key of B minor increases understanding in music theory concepts, and applications.

II: Getting to Know The Possibilities:

1. B minor Diatonic Scale
2. B minor Pentatonic Scale
3. B minor Arpeggio Scale
4. B minor Pattern / Structure Progressions

III: Exploring Concepts:

1. B minor Diatonic Scale: 2 Sharps.
   B C#° D E F# G A
2. B minor Pentatonic Scale: 1 Sharp.
   B D E F# A
3. B minor Arpeggio Scale: 3 Notes, 1 Sharp.
   B D F#
4. B minor Diatonic Chords:
   Bm C#° DM Em F#m G^M A^M
5. B minor (Major Chords):
   D G A
6. B minor (Minor Chords):
   B E F#
7. B minor (Diminish Chord):
   C#°

IV: Possibile Solutions:

B minor "Minor" Chord Progressions

1. Bm	Em	F#m
2. Em	F#m	Em
3. Em	Bm	F#m
4. Bm	F#m	Em
5. Em	Bm	F#m
6. Bm	F#m	Em

B minor "Major" Chord Progressions

1. G^M	C^M	D^M
2. G^M	D^M	C^M
3. C^M	D^M	G^M
4. C^M	G^M	D^M
5. G^M	C^M	D^M
6. G^M	D^M	C^M

B minor "Diminish Chord Progressions

1. Bm	C#°	Bm
2. Em	C#°	Bm
3. Bm	C#°	Em
4. F#^M	C#°	Bm
5. Bm	C#°	D^M

V: Testing, and Evaluation:

B minor / D Major Chord Progressions

1. F#m	D^M	G^M	A^M
2. Bm	G^M	A^M	D^M
3. Em	A^M	D^M	F#
4. G^M	A^M	D^M	Bm
5. G^M	C^M	Am	D^M
6. D^M	G^M	Bm	A^M
7. Bm	G^M	C#°	D^M
8. Em	D^M	C#°	Bm

Discovery Design Brief:
11.01-31-24 / AP
STEM In Education
Principles of Engineering
Teaching, and Learning Concepts
"STEM In Music"
Key of A Major (Notes, and Chords)

I: Possibility Statement:

Exploring the key of A Major increases understanding in music theory concepts, and applications.

II: Getting to Know The Possibilities:

1. A Major Diatonic Scale
2. A Major Pentatonic Scale
3. A Major Arpeggio Scale
4. A Major Pattern / Structure Progressions

III: Exploring Concepts:

1. A Major Diatonic Scale: 3 Sharps.
   A B C# D E F# G#
2. A Major Pentatonic Scale: 5 Notes,3 Sharps.
   A B C# E F#
3. A Major Arpeggio Scale: 3 Notes, 1 Sharp.
   A C# E
4. A Major Diatonic Chords:
   A^M B^m C#^m D^M E^M F#^m G#°
5. A Major (Major Chords):
   A^M D^M E^M
6. A Major (Minor Chords):
   B C#m F#m

7. A Major (Diminish Chord):
   G#°

IV: Possibile Solutions:

A Major "Major" Chord Progressions

1. A^M	D^M	E^M
2. A^M	E^M	D^M
3. D^M	E^M	A^M
4. D^M	A^M	E^M
5. E^M	A^M	D^M
6. E^M	D^M	A^M

A Major "Minor" Chord Progressions

1. F#m	Bm	C#m
2. F#m	C#m	Bm
3. Bm	C#m	F#m
4. Bm	F#m	C#m
5. C#m	Bm	F#m
6. F#m	C#m	Bm

A Major "Diminish Chord Progressions

1. A^M	G#°	A^M
2. D^M	G#°	E^M
3. E^M	G#°	E^M
4. E^M	G#°	A^M
5. A^M	G#°	F#m

V: Testing, and Evaluation:

A Major / F# minor Chord Progressions

1. A^M	Cm	D^M	E^M
2. F#m	D^M	A^M	E^M
3. A^M	E^M	A^M	F#m
4. D^M	E^M	A^M	Cm
5. E^M	A^M	Cm	D^M
6. A^M	E^M	Cm	E^M
7. D^M	A^M	G#°	E^M
8. F#m	A^M	G#°	D^M

Discovery Design Brief:
12.01-31-24 / AP
STEM In Education
Principles of Engineering
Teaching, and Learning Concepts
"STEM In Music"
Key of F# minor (Notes, and Chords)

I: Possibility Statement:

Exploring the key of F# minor increases understanding in music theory concepts, and applications.

II: Getting to Know The Possibilities:

1. F# minor Diatonic Scale
2. F# minor Pentatonic Scale
3. F# minor Arpeggio Scale
4. F minor Pattern / Structure Progressions

III: Exploring Concepts:

1. F# minor Diatonic Scale: 7 Notes, 3 Sharps.
   F# G#° A B C# D E
2. F# minor Pentatonic Scale: 5 Notes, 2 Sharps.
   F# A B C# E
3. F# minor Arpeggio Scale: 3 Notes, 1 Flat.
   F# Eb G
4. F# minor Diatonic Chords:
   F#m G#° AM Bm Cm D^M E^M
5. F# minor (Major Chords):
   A B D
6. F# minor (Minor Chords):
   F# B C#
7. F# minor (Diminish Chord):
   G#°

IV: Possibile Solutions:

F# minor "Minor" Chord Progressions

1. F#m	Bm	C#m
2. F#m	C#m	Bm
3. Bm	F#m	C#m
4. Bm	C#m	F#m
5. C#m	F#m	Bm
6. C#m	Bm	F#m

F#minor "Major" Chord Progressions

1. A^M	D^M	E^M
2. D^M	E^M	A^M
3. A^M	D^M	D^M
4. D^M	A^M	E^M
5. E^M	A^M	D^M
6. D^M	A^M	E^M

F# minor "Diminish Chord Progressions

1. F#m	G°	F#m
2. Bm	G°	F#m
3. C#m	G°	F#m
4. A^M	G°	F#mm
5.F#m	G°	A^M

V: Testing, and Evaluation:

F# minor / A Major Chord Progressions

1.F#m	A^M	D^M	E^M
2. C#m	D^M	E^M	A^M
3. F#m	B^M	A^M	Gm
4. A^M	D^M	E^M	C#m
5. A^M	D^M	F#m	E^M
6. E^M	A^M	C#m	A^M
7. C#m	A^M	G#°	E^M
8. F#m	Eb^M	G#°	C#m

Discovery Design Brief:
13.01-31-24 / AP
STEM In Education
Principles of Engineering
Teaching, and Learning Concepts
"STEM In Music"
Key of Eb Major (Notes, and Chords)

I: Possibility Statement:

Exploring the key of Eb Major increases understanding in music theory concepts, and applications.

II: Getting to Know The Possibilities:

1. Eb Major Diatonic Scale
2. Eb Major Pentatonic Scale
3. Eb Major Arpeggio Scale
4. Eb Major Pattern / Structure Progressions

III: Exploring Concepts:

1. Eb Major Diatonic Scale: 3 Flats.
   Eb F G Ab Bb C D
2. Eb Major Pentatonic Scale: 5 Notes, 2 Flats.
   Eb F G Bb C
3. Eb Major Arpeggio Scale: 3 Notes, 2 Flats.
   Eb G Bb
4. Eb Major Diatonic Chords:
   Eb^M F^m G^m Ab^M Bb^M C^m D°
5. Eb Major (Major Chords):
   Eb^M Ab^M Bb^M
6. Eb Major (Minor Chords):
   Cm Fm Gm

7. Eb Major (Diminish Chord):
   D°

IV: Possibile Solutions:

Eb Major "Major" Chord Progressions

1. Eb^M	Ab^M	Bb^M
2. Ab^M	Bb^M	Eb^M
3. Bb^M	Eb^M	Ab^M
4. Eb^M	Bb^M	Ab^M
5. Ab^M	Eb^M	Bb^M
6. Bb^M	Ab^M	Eb^M

Eb Major "Minor" Chord Progressions

1. Cm	Fm	Gm
2. Cm	Gm	Fm
3. Gm	Fm	Cm
4. Gm	Cm	Fm
5. Fm	Gm	Cm
6. Fm	Cm	Gm

Eb Major "Diminish Chord Progressions

1. Eb^M	C#°	Eb^M
2. Ab^M	C#°	Eb^M
3. Bb^M	C#°	Eb^M
4. Eb^M	C#°	Bb^M
5. Ab^M	C#°	Bb^M

V: Testing, and Evaluation:

Eb Major / C minor Chord Progressions

1. Eb^M	Cm	Ab^M	Bb^M
2. Cm	Bb^M	Ab^M	Eb^M
3. Ab^M	Eb^M	Bb^M	Fm
4. Bb^M	AB^M	Eb^M	Gm
5. Eb^M	Ab^M	Cm	Eb^M
6. Ab^M	Bb^M	Gm	Eb^M
7. Bb^M	Gb^M	D°	Eb^M
8. Fm	Ab^M	D°	Bb^M

Discovery Design Brief:
14.01-31-24 / AP
STEM In Education
Principles of Engineering
Teaching, and Learning Concepts
"STEM In Music"
Key of C minor (Notes, and Chords)

I: Possibility Statement:
Exploring the key of C minor increases understanding in music theory concepts, and applications.

II: Getting to Know The Possibilities:
1. C minor Diatonic Scale
2. C minor Pentatonic Scale
3. C minor Arpeggio Scale
4. C minor Pattern / Structure Progressions

III: Exploring Concepts:
1. C minor Diatonic Scale: 7 Notes, 3 Flats.
   C D° Eb F G Ab Bb
2. C minor Pentatonic Scale: 5 Notes, 2 Flats.
   C Eb F G Bb
3. C minor Arpeggio Scale: 3 Notes, 1 Flat.
   C Eb G
4. C minor Diatonic Chords:
   Cm D° EbM Fm Gm Ab^M Bb^M
5. C minor (Major Chords):
   Eb Ab Bb
6. C minor (Minor Chords):
   C F G
7. C minor (Diminish Chord):
   D°

IV: Possibile Solutions:

C minor "Minor" Chord Progressions

1. Cm	Fm	Gm
2. Fm	Cm	Gm
3. Gm	Cm	Fm
4. Cm	Gm	Fm
5. Cm	Gm	Fm
6. Fm	Gm	Cm

C minor "Major" Chord Progressions

1. Ab^M	Bb^M	Eb^M
2. Bb^M	Eb^M	Ab^M
3. Eb^M	Bb^M	Ab^M
4. Ab^M	Eb^M	Bb^M
5. Bb^M	Ab^M	Eb^M
6. Eb^M	Ab^M	Bb^M

C minor "Diminish Chord Progressions

1. Cm	D°	Cm
2. Fm	D°	Cm
3. Gm	D°	Cm
4. Eb^M	D°	Cm
5. Cm	D°	Eb^M

V: Testing, and Evaluation:

C minor / Eb Major Chord Progressions

1. Gm	Eb^M	Ab^M	Bb^M
2. Cm	Ab^M	Bb^M	Eb^M
3. Fm	Bb^M	Eb^M	Gm
4. Ab^M	Bb^M	Eb^M	Cm
5. Ab^M	Bb^M	Gm	Eb^M
6. Eb^M	Ab^M	Cm	Bb^M
7. Cm	Ab^M	D°	Eb^M
8. Fm	Eb^M	D°	Cm

<u>Discovery Design Brief:</u>
<u>15.01-31-24 / AP</u>
STEM In Education
Principles of Engineering
Teaching, and Learning Concepts
<u>"STEM In Music"</u>
<u>Key of Ab Major</u> (Notes, and Chords)

<u>I: Possibility Statement:</u>

Exploring the key of Ab Major increases understanding in music theory concepts, and applications.

<u>II: Getting to Know The Possibilities:</u>

1. Ab Major Diatonic Scale
2. Ab Major Pentatonic Scale
3. Ab Major Arpeggio Scale
4. Ab Major Pattern / Structure Progressions

<u>III: Exploring Concepts:</u>

1. Ab Major Diatonic Scale: 7 Notes, 4 Flats.
   Ab Bb C Db Eb F G
2. Ab Major Pentatonic Scale: 3 Flats.
   Ab Bb C Eb F
3. Ab Major Arpeggio Scale: 3 Notes, 2 Flats.
   Ab C Eb
4. Ab Major Diatonic Chords:
   Ab^M Bb^m C^m D^M Eb^M F^m G°
5. Ab Major (Major Chords):
   Ab^M D^M Eb^M
6. Ab Major (Minor Chords):
   Bbm Cm Fm

7. Ab Major (Diminish Chord):
   G°

IV: Possibile Solutions:

Ab Major "Major" Chord Progressions

1. Ab^M	Db^M	Eb^M
2. Db^M	Eb^M	Ab^M
3. Eb^M	Db^M	Eb^M
4. Db^M	Ab^M	Eb^M
5. Ab^M	Eb^M	Bb^M
6. Eb^M	Db^M	Ab^M

Ab Major "Minor" Chord Progressions

1. Bbm	Cm	Fm
2. Fm	Cm	Bbm
3. Cm	Bam	Fm
4. Bbm	Fm	Cm
5. Fm	Bbm	Cm
6. Cm	Fm	Bbm

Ab Major "Diminish Chord Progressions

1. Ab^M	G°	Ab^M
2. Db^M	G°	Ab^M
3. Eb^M	G°	Ab^M
4. Db^M	G°	Db^M
5. Ab^M	G°	F^m

V: Testing, and Evaluation:

Ab Major / F minor Chord Progressions

1. Ab^M	Fm	Eb^M	Db^M
2. Fm	Eb^M	Db^M	Ab^M
3. Ab^M	Eb^M	Bb^M	Fm
4. Db^M	Ab^M	Eb^M	Fm
5. Eb^M	Ab^M	Fm	Bb^M
6. Ab^M	Db^M	Abm	Eb^M
7. Bb^M	Db^M	G°	Eb^M
8. Fm	Ab^M	G°	Bb^M

Discovery Design Brief:
16.01-31-24 / AP
STEM In Education
Principles of Engineering
Teaching, and Learning Concepts
"STEM In Music"
Key of F minor (Notes, and Chords)

I: Possibility Statement:
Exploring the key of F minor increases understanding in music theory concepts, and applications.

II: Getting to Know The Possibilities:
1. F minor Diatonic Scale
2. F minor Pentatonic Scale
3. F minor Arpeggio Scale
4. F minor Pattern / Structure Progressions

III: Exploring Concepts:
1. F minor Diatonic Scale: 7 Notes, 4 Flats.
   F G Ab Bb C Db Eb
2. F minor Pentatonic Scale: 5 Notes, 3 Flats.
   F Ab Bb C Eb
3. F minor Arpeggio Scale: 3 Notes, 1 Flat.
   F Ab C
4. F minor Diatonic Chords:
   Fm G° AbM Bbm Cm Db^M Eb^M
5. F minor (Major Chords):
   Ab Db Eb
6. F minor (Minor Chords):
   F Bb C
7. F minor (Diminish Chord):
   G°

IV: Possibile Solutions:

F minor "Minor" Chord Progressions

1. Cm	Fm	Bbm
2. Fm	Fm	Bm
3. Bbm	Cm	Fm
4. Cm	Bbm	Fm
5. Bbm	Fm	Cm
6. Fm	Bbm	Cm

F minor "Major" Chord Progressions

1. Ab^M	Db^M	Eb^M
2. Db^M	Eb^M	Ab^M
3. Eb^M	Db^M	Ab^M
4. Ab^M	Eb^M	Db^M
5. Db^M	Ab^M	Eb^M
6. Eb^M	Ab^M	Db^M

F minor "Diminish Chord Progressions

1. Cm	G°	Cm
2. Fm	G°	Cm
3. Bbm	G°	Cm
4. Ab^M	G°	Cm
5. Fm	G°	Ab^M

V: Testing, and Evaluation:

F minor / Ab Major Chord Progressions

1. Gm	Eb^M	Ab^M	Bb^M
2. Cm	Ab^M	Bb^M	Eb^M
3. Fm	Bb^M	Eb^M	Gm
4. Ab^M	Bb^M	Eb^M	Cm
5. Ab^M	Bb^M	Gm	Eb^M
6. Eb^M	Ab^M	Cm	Bb^M
7. Cm	Ab^M	D°	Eb^M
8. Fm	Eb^M	D°	Cm

<u>Discovery Design Brief:</u>
<u>17.01-31-24 / AP</u>
STEM In Education
Principles of Engineering
Teaching, and Learning Concepts
<u>"STEM In Music"</u>
<u>Key of E Major</u> (Notes, and Chords)

<u>I: Possibility Statement:</u>

Exploring the key of E Major increases understanding in music theory concepts, and applications.

<u>II: Getting to Know The Possibilities:</u>

1. E Major Diatonic Scale
2. E Major Pentatonic Scale
3. E Major Arpeggio Scale
4. E Major Pattern / Structure Progressions

<u>III: Exploring Concepts:</u>

1. E Major Diatonic Scale: 7 Notes, 4 Flats.
   E F# G# A B C# D#
2. E Major Pentatonic Scale: 3 Sharp.
   E F# G# B C#
3. E Major Arpeggio Scale: 3 Notes, 1 Sharp.
   E G# B
4. E Major Diatonic Chords:
   E^M F#^m G#^m A^M B^M C#^m D#°
5. E Major (Major Chords):
   E^M A^M B^M
6. E Major (Minor Chords):
   F#m G#m C#m

7. E Major (Diminish Chord):
   D#°

IV: Possibile Solutions:

E Major "Major" Chord Progressions

1. E^M	A^M	B^M
2. E^M	B^M	A^M
3. B^M	E^M	A^M
4. B^M	A^M	E^M
5. E^M	B^M	A^M
6. E^M	A^M	B^M

E Major "Minor" Chord Progressions

1. C#m	F#m	G#m
2. C#m	G#m	F#m
3. F#m	C#m	G#m
4. F#m	G#m	C#m
5. G#m	C#m	F#m
6. G#m	F#m	C#m

E Major "Diminish Chord Progressions

1. E^M	D#°	E^M
2. A^M	D#°	E^M
3. B^M	D#°	E^M
4. C#m	D#°	E^M
5. E^M	D#°	C#^m

V: Testing, and Evaluation:

E Major / C# minor Chord Progressions

1. E^M	C#m	B^M	A^M
2. F#m	B^M	A^M	E^M
3. B^M	A^M	B^M	F#m
4. A^M	E^M	B^M	C#
5. B^M	E^M	G#m	B^M
6. A^M	D^M	G#m	E^M
7. B^M	D^M	D#°	E^M
8. F#m	A^M	D#°	E^M

<u>Discovery Design Brief:</u>
<u>18.01-31-24 / AP</u>
STEM In Education
Principles of Engineering
Teaching, and Learning Concepts
<u>"STEM In Music"</u>
<u>Key of C# minor</u> (Notes, and Chords)

<u>I: Possibility Statement:</u>

Exploring the key of C# minor increases understanding in music theory concepts, and applications.

<u>II: Getting to Know The Possibilities:</u>

1. C# minor Diatonic Scale
2. C# minor Pentatonic Scale
3. C# minor Arpeggio Scale
4. C# minor Pattern / Structure Progressions

<u>III: Exploring Concepts:</u>

1. C# minor Diatonic Scale: 7 Notes, 4 Sharps.
   C# D#° E F# G# A B
2. C# minor Pentatonic Scale: 5 Notes, 4 Sharps.
   C# E F# G# D#
3. C# minor Arpeggio Scale: 3 Notes, 2 Sharps.
   C# E G#
4. C# minor Diatonic Chords:
   C#m D#° E^M F#m G#m A^M B^M
5. C minor (Major Chords):
   Eb Ab Bb
6. C# minor (Minor Chords):
   C# F# G#
7. C# minor (Diminish Chord):
   D#°

IV: Possibile Solutions:

C# minor “Minor” Chord Progressions

1. C#m	F#m	G#m
2. F#m	F#m	G#m
3. G#m	C#m	F#m
4. C#m	G#m	F#m
5. G#m	F#m	C#m
6. F#m	G#m	C#m

C# minor “Major” Chord Progressions

1. A^M	B^M	E^M
2. B^M	E^M	A^M
3. E^M	B^M	A^M
4. A^M	E^M	B^M
5. B^M	A^M	E^M
6. E^M	A^M	B^M

C# minor “Diminish Chord Progressions

1. C#m	D°	C#m
2. F#m	D°	C#m
3. G#m	D°	C#m
4. E^M	D°	C#m
5. C#m	D°	E^M

V: Testing, and Evaluation:

C# minor / E Major Chord Progressions

1. G#m	E^M	A^M	B^M
2. C#m	A^M	B^M	E^M
3. F#m	B^M	E^M	G#m
4. A^M	B^M	Eb^M	Cm
5. A^M	B^M	G#m	E^M
6. E^M	A^M	C#m	B^M
7. C#m	A^M	D#°	E^M
8. F#m	E^M	D#°	C#m

Discovery Design Brief:
21.01-31-24 / AP
STEM In Education
Principles of Engineering
Teaching, and Learning Concepts
"STEM In Music"
Key of Db Major (Notes, and Chords)

I: Possibility Statement:
Exploring the key of Db Major increases understanding in music theory concepts, and applications.

II: Getting to Know The Possibilities:
1. Db Major Diatonic Scale
2. Db Major Pentatonic Scale
3. Db Major Arpeggio Scale
4. Db Major Pattern / Structure Progressions

III: Exploring Concepts:
1. Db Major Diatonic Scale: 7 Notes, 5 Flats.
   Db Eb F Gb Ab Bb C
2. Db Major Pentatonic Scale: 4 Sharps.
   Db Eb F Ab Bb
3. Db Major Arpeggio Scale: 3 Notes, 1 Sharp.
   Db F Ab
4. Db Major Diatonic Chords:
   Db^M Eb^m F^m Gb^M Ab^M Bb^m Cb
5. Db Major (Major Chords):
   Db^M Gb^M Ab^M
6. Db Major (Minor Chords):
   Ebm Fm Bbm

7. Db Major (Diminish Chord):
   C°

IV: Possibile Solutions:

Db Major "Major" Chord Progressions

1. Db^M	Gb^M	Ab^M
2. Db^M	Ab^M	Gb^M
3. Ab^M	Gb^M	Db^M
4. Ab^M	Db^M	Gb^M
5. Gb^M	Ab^M	Db^M
6. Gb^M	Db^M	Ab^M

Db Major "Minor" Chord Progressions

1. Bbm	Ebm	Fm
2. Bbm	Fm	Ebb
3. Fm	Ebm	Bbm
4. Fm	Bbm	Fm
5. Ebm	Fm	Bbm
6. Ebm	Bbm	Fm

Db Major "Diminish Chord Progressions

1. Db^M	C°	Db^M
2. Ab^M	C°	Db^M
3. Gb^M	C°	Db^M
4. Fm	C°	Db^M
5. Db^M	C°	F^m

V: Testing, and Evaluation:

Db Major / Bb minor Chord Progressions

1. Db^M	Bbm	Ab^M	Gb^M
2. Fm	Gb^M	Db^M	Ab^M
3. Gb^M	Ab^M	Db^M	Fm
4. Ab^M	Eb^M	Gb^M	Ebm
5. B^M	Gb^M	Bbm	Db^M
6. Ab^M	Db^M	Ebm	Gb^M
7. Gb^M	Gb^M	C°	Db^M
8. Fm	Ab^M	C°	Db^M

Discovery Design Brief:
22.01-31-24 / AP
STEM In Education
Principles of Engineering
Teaching, and Learning Concepts
"STEM In Music"
Key of Bb minor (Notes, and Chords)

I: Possibility Statement:

Exploring the key of Bb minor increases understanding in music theory concepts, and applications.

II: Getting to Know The Possibilities:

1. Bb minor Diatonic Scale
2. Bb minor Pentatonic Scale
3. Bb minor Arpeggio Scale
4. Bb minor Pattern / Structure Progressions

III: Exploring Concepts:

1. Bb minor Diatonic Scale: 7 Notes, 5 Flats.
   Bb C Db Eb F Gb Ab
2. Bb minor Pentatonic Scale: 5 Notes, 4 Flates.
   Bb Db Eb F Ab
3. Bb minor Arpeggio Scale: 3 Notes, 2 Flats.
   Bb Db F
4. Bb minor Diatonic Chords:
   Bbm C° Db^M Ebm Fm Gb^M Ab^M
5. Bb minor (Major Chords):
   Db Gb Ab
6. Bb minor (Minor Chords):
   Bb Eb F
7. Bb minor (Diminish Chord):
   C°

IV: Possibile Solutions:

Bb minor "Minor" Chord Progressions

1. Bbm	Ebm	Fm
2. Bbm	Fm	Ebm
3. Fm	Ebm	Bbm
4. Fm	Bbm	Ebm
5. Ebm	Fm	Bbm
6. Ebm	Bbm	Fm

Bb minor "Major" Chord Progressions

1. Ab^M	Db^M	Gb^M
2. Ab^M	Gb^M	Db^M
3. Gb^M	Db^M	Ab^M
4. Gb^M	Ab^M	Db^M
5. Db^M	Ab^M	Gb^M
6. Db^M	Gb^M	Ab^M

Bb minor "Diminish Chord Progressions

1. Bbm	C°	Bbm
2. Gbm	C°	Bbm
3. Fm	C°	Bbm
4. Ab^M	C°	F#m
5. F#m	C°	Gb^M

V: Testing, and Evaluation:

Bb minor / Db Major Chord Progressions

1. Bbm	Db^M	Gb^M	Ab^M
2. Ebm	Ab^M	Db^M	Gb^M
3. B#m	Gb^M	D^M	Fm
4. Ab^M	Db^M	Gb^M	Ebm
5. Ab^M	Gb^M	Fm	Db^M
6. Db^M	Ab^M	Bbm	Gb^M
7. Fm	Ab^M	C°	Db^M
8. Ebm	Gb^M	C°	Fm

STEM Educational Enrichment Services LLC

"STEM In Music"

Principles of Engineering
Teaching, and Learning

Grades K - 6

Pre / Post Surveys

TMS STEM Studios

Instructor / Facilitator: Mr. Michael A. Johnson

Principles of Engineering

Pre / Post-Survey

Grades 1 & 2

*Direction: This is STEM Studios Principles of Engineering survey. Please circle **True** if you are sure the answer is yes, circle **False** if you are sure the answer is no, or circle **I Don't Know** if you're not sure of the answer.*

Science:

1. Chemical notation for water is H_2O.

True ***False*** ***I Don't Know***

2. Chemical notation for oxygen in Earth's Troposphere is O_2.

True ***False*** ***I Don't Know***

3. Chemical notation for nitrogen in Earth's Troposphere is N_2.

True ***False*** ***I Don't Know***

Technology:

4. Coding is used for programming integrated circuits, and systems.

True ***False*** ***I Don't Know***

5. Coding is used for computer processing.

True ***False*** ***I Don't Know***

6. Coding is used for computer programing applications.

True ***False*** ***I Don't Know***

Engineering:

7. A lever is a mechanism.

True ***False*** ***I Don't Know***

8. A lever has a fulcrum.

True ***False*** ***I Don't Know***

9. A lever has an effort force.

True ***False*** ***I Don't Know***

10. A lever has a load force.

True ***False*** ***I Don't Know***

Mathematics:

11. Mass of an object is measured in grams.

True ***False*** ***I Don't Know***

12. Mass of an object is measured in meters.

True ***False*** ***I Don't Know***

13. Mass of an object is measured in liters.

True ***False*** ***I Don't Know***

TMS STEM Studios

Instructor / Facilitator: Mr. Michael A. Johnson
Principles of Engineering
Pre / Post-Survey
Grades 3 & 4

*Direction: This is STEM Studios Principles of Engineering survey. Please circle **True** if you are sure the answer is yes, circle **False** if you are sure the answer is no, or circle **I Don't Know** if you're not sure of the answer.*

Science:

1. Compression is a pushing force acting on a static, or dynamic system.

True ***False*** ***I Don't Know***

2. Compression is a pulling force acting on a static, or dynamic system.

True ***False*** ***I Don't Know***

3. Tension is a pushing force acting on a static, or dynamic system.

True ***False*** ***I Don't Know***

4. Tension is a pulling force acting on a static, or dynamic system.

True ***False*** ***I Don't Know***

Technology:

5. Makey Makey is a computer software application.

True ***False*** ***I Don't Know***

6. Makey Makey is a computer hardware device.

True ***False*** ***I Don't Know***

7. Arduino Uno uses C / C++ Processing.

True ***False*** ***I Don't Know***

8. Arduino Software IDE is used for coding.

True ***False*** ***I Don't Know***

9. Techno K'nex uses C / C++ Process.

True ***False*** ***I Don't Know***

10. Techno K'nex uses a simulated controlled interface.

True ***False*** ***I Don't Know***

11. Techno K'nex communicates using a Leonardo interface.

True ***False*** ***I Don't Know***

Engineering:

12. Force normal creates gravitational pull.

True ***False*** ***I Don't Know***

13. Force normal pushes against gravity.

True ***False*** ***I Don't Know***

14. Force normal has no net force.

True ***False*** ***I Don't Know***

15. Force normal has a net force.

True ***False*** ***I Don't Know***

Mathematics:

16. 10.0cm = 100m

True ***False*** ***I Don't Know***

17. 10.0cm = 0.10m

True ***False*** ***I Don't Know***

18. 10cm = 0.01m

True ***False*** ***I Don't Know***

19. 5.0m = 0.500cm

True ***False*** ***I Don't Know***

20. 5.0m = 500.0cm

True ***False*** ***I Don't Know***

21. 5.0m = 5000.0cm

True ***False*** ***I Don't Know***

TMS STEM Studios

Instructor / Facilitator: Mr. Michael A. Johnson
Principles of Engineering
Pre / Post-Survey
Grades 5th & 6th

*Direction: This is STEM Studios Principles of Engineering survey. Please circle **True** if you are sure the answer is yes, circle **False** if you are sure the answer is no, or circle **I Don't Know** if you're not sure of the answer.*

Science:

1. Gravitational force occurs on objects acting outside of a static system.

True ***False*** ***I Don't Know***

2. Gravitational force occurs on objects acting inside of a static system.

True ***False*** ***I Don't Know***

3. Normal force (Fn) occurs on objects acting outside of a static system.

True ***False*** ***I Don't Know***

4. Normal force (Fn) occurs on objects acting inside of a static system.

True ***False*** ***I Don't Know***

Technology:

5. Makey Makey earth technological system communicates with positive battery terminal.

True ***False*** ***I Don't Know***

6. Makey Makey earth technological system communicates with ground battery terminal.

True ***False*** ***I Don't Know***

7. Arduino Uno uses C / C++ Processing.

True ***False*** ***I Don't Know***

6. Arduino Software IDE is used for coding.

True ***False*** ***I Don't Know***

7. Techno K'nex uses C / C++ Process.

True ***False*** ***I Don't Know***

8. Techno K'nex uses a simulated controlled Leonardo interface.

True ***False*** ***I Don't Know***

9. Techno K'nex communicates using a Leonardo interface.

True ***False*** ***I Don't Know***

Engineering:

10. External forces exist outside of a static system.

True ***False*** ***I Don't Know***

11. Internal forces exist outside of a static system.

True ***False*** ***I Don't Know***

12. Mechanical stress exist in dynamic systems.

True ***False*** ***I Don't Know***

13. Mechanical stress exist in static systems

True ***False*** ***I Don't Know***

Mathematics:

14. Force is measured in Newtons

True ***False*** ***I Don't Know***

15. Force is measured in Volume.

True ***False*** ***I Don't Know***

16. Force is measured in Meters / Seconds.

True ***False*** ***I Don't Know***

STEM Educational Enrichment Services LLC.

"STEM In Music"

Principles of Engineering
Teaching, and Learning Concepts

Grades 7th - 12th

Pre / Post Surveys

STEMedu Studios Live

Instructor / Facilitator: Mr. Michael A. Johnson
Principles of Engineering
Pre / Post-Survey
Grades 7th - 12th

Direction: This is STEM Studios Principles of Engineering survey. Please circle ***True*** *if you are sure the answer is yes, circle* ***False*** *if you are sure the answer is no, or circle* ***I Don't Know*** *if you're not sure of the answer.*

Music Theory, Concepts, and Applications:

1. Diatonic Scale has 3 notes.

True ***False*** ***I Don't Know***

2. Diatonic Scale has 7 notes.

True ***False*** ***I Don't Know***

3. Diatonic Scale has 5 notes.

True ***False*** ***I Don't Know***

4. Pentatonic Scale is composed of 3 notes.

True ***False*** ***I Don't Know***

5. Pentatonic Scale is composed of 7 notes.

True ***False*** ***I Don't Know***

6. Pentatonic Scale is composed of 5 notes.

True ***False*** ***I Don't Know***

7. Arpeggio Scale contains 3 notes.

True ***False*** ***I Don't Know***

8. Arpeggio Scale contains 7 notes.

True ***False*** ***I Don't Know***

9. Arpeggio Scale contains 5 notes.

True ***False*** ***I Don't Know***

10. Each Key Signature consist of 3 notes.

True ***False*** ***I Don't Know***

11. Each Key Signature consist of 7 notes.

True ***False*** ***I Don't Know***

12. Each Key Signature consist of 5 notes.

True ***False*** ***I Don't Know***

13. Each Key consist of 5 Major Chords.

True ***False*** ***I Don't Know***

14. Each Key consist of 3 Major Chords.

True ***False*** ***I Don't Know***

15. Each Key consist of 5 Major Chords.

True ***False*** ***I Don't Know***

16. Each Key consist of 7 Major Chords.

True ***False*** ***I Don't Know***

17. Each Key consist of 3 Minor Chords.

True ***False*** ***I Don't Know***

18. Each Key consist of 5 Minor Chords.

True ***False*** ***I Don't Know***

19. Each Key consist of 7 Minor Chords.

True ***False*** ***I Don't Know***

20. Each Key consist of 3 diminish Chords.

True ***False*** ***I Don't Know***

21. Each Key consist of 1 diminish Chords.

True ***False*** ***I Don't Know***

22. Each Key consist of 5 diminish Chords.

True ***False*** ***I Don't Know***

23. Key of C contains no Flats, and no Sharps.

True ***False*** ***I Don't Know***

24. Key of C contains 1 Flat, and no Sharps.

True ***False*** ***I Don't Know***

25. Key of C contains no Flats, and 1 Sharp.

26. Key of Am contains no Flats, and no Sharps.

True ***False*** ***I Don't Know***

27. Key of Am contains no Flats, and no Sharps.

True ***False*** ***I Don't Know***

28. Key of Am contains no Flats, and no Sharps.

True ***False*** ***I Don't Know***

29. Key of G^M contains no Flats, and no Sharps.

True ***False*** ***I Don't Know***

30. Key of Am contains no Flats, and 1 Sharp.

True ***False*** ***I Don't Know***

31. Key of Am contains 1 Flat, and no Sharps.

True ***False*** ***I Don't Know***

32. Key of G^M contains no Flats, and no Sharps.

True ***False*** ***I Don't Know***

32. Key of G^M contains 1 Flat, and no Sharps.

True ***False*** ***I Don't Know***

31. Key of G^M contains no Flats, and 1 Sharps.

True ***False*** ***I Don't Know***

32. Key of Em contains no Flats, and no Sharps.

True ***False*** ***I Don't Know***

33. Key of Em contains 1 Flat, and no Sharps.

True ***False*** ***I Don't Know***

34. Key of Em contains no Flat, and 1 Sharps.

True ***False*** ***I Don't Know***

35. Key of F^M contains no Flats, and no Sharps.

True ***False*** ***I Don't Know***

36. Key of F^M contains 1 Flat, and 1 Sharps.

True ***False*** ***I Don't Know***

37. Key of F^M contains 1 Flat, and no Sharps.

True ***False*** ***I Don't Know***

38. Key of Dm contains no Flats, and no Sharps.

True ***False*** ***I Don't Know***

39. Key of Dm contains 1 Flat, and no Sharps.

True ***False*** ***I Don't Know***

40. Key of Dm contains no Flat, and 1 Sharps.

True ***False*** ***I Don't Know***

41. Key of D^M contains 3 Flats, and no Sharps.

True ***False*** ***I Don't Know***

42. Key of D^M contains 1 Flat, and 2 Sharps.

True ***False*** ***I Don't Know***

43. Key of D^M contains no Flats, and 2 Sharps.

True ***False*** ***I Don't Know***

44. Key of Bm contains 2 Flats, and no Sharps.

True ***False*** ***I Don't Know***

45. Key of Bm contains 1 Flat, and no Sharps.

True ***False*** ***I Don't Know***

46. Key of Bm contains no Flat, and 2 Sharps.

True ***False*** ***I Don't Know***

47. Key of Bb^M contains 2 Flats, and no Sharps.

True ***False*** ***I Don't Know***

48. Key of Bb^M contains 1 Flat, and 2 Sharps.

True ***False*** ***I Don't Know***

49. Key of Bb^M contains no Flats, and 2 Sharps.

True ***False*** ***I Don't Know***

50. Key of Gm contains 2 Flats, and no Sharps.

True ***False*** ***I Don't Know***

51. Key of Gm contains 1 Flat, and no Sharps.

True ***False*** ***I Don't Know***

52. Key of Gm contains no Flat, and 2 Sharps.

True ***False*** ***I Don't Know***

53. Key of A^M contains no Flats, and 1 Sharps.

True ***False*** ***I Don't Know***

54. Key of A^M contains 1 Flat, and 2 Sharps.

True ***False*** ***I Don't Know***

55. Key of A^M contains no Flats, and 3 Sharps.

True ***False*** ***I Don't Know***

56. Key of F#m contains no Flats, and 3 Sharps.

True ***False*** ***I Don't Know***

57. Key of F#m contains 1 Flat, and no Sharps.

True ***False*** ***I Don't Know***

58. Key of F#m contains no Flat, and 2 Sharps.

True ***False*** ***I Don't Know***

59. Key of Eb^M contains no Flats, and 3 Sharps.

True ***False*** ***I Don't Know***

60. Key of Eb^M contains 1 Flat, and 2 Sharps.

True ***False*** ***I Don't Know***

61. Key of Eb^M contains 3 Flats, and no Sharps.

True ***False*** ***I Don't Know***

62. Key of Gm contains no Flats, and 3 Sharps.

True ***False*** ***I Don't Know***

63. Key of Gm contains 1 Flat, and no Sharps.

True ***False*** ***I Don't Know***

64. Key of Gm contains no Flat, and 2 Sharps.

True ***False*** ***I Don't Know***

65. Key of E^M contains no Flats, and 3 Sharps.

True ***False*** ***I Don't Know***

66. Key of E^M contains no Flats, and 4 Sharps.

True ***False*** ***I Don't Know***

67. Key of E^M contains no Flats, and 4 Sharps.

True ***False*** ***I Don't Know***

68. Key of C#m contains no Flats, and 3 Sharps.

True ***False*** ***I Don't Know***

69. Key of C#m contains 1 Flat, and no Sharps.

True ***False*** ***I Don't Know***

70. Key of C#m contains no Flat, and 4 Sharps.

True ***False*** ***I Don't Know***

Special Thanks:

Special thanks to *My Legendary Father James Pop Johnson. Mother: Laurestene M. Johnson-Smith, and husband Larry Smith.* ***Daughter:*** *Jasmine,* ***J5:*** Tim Sr, Mary, Tim Jr, Israel, and Jade, ***Brothers:*** James, Lawrence (RIP), Kevin, and Timothy. ***Sisters:*** Yvonne, and Juanita. ***Sister in Law:*** Mary ***Aunts:*** Wanda, Sue, Gerri (RIP), and Elizabeth (RIP). ***Uncle*** Harold, and Isaiah (RIP). ***Nephews:*** Emato, Lawrence, Andreas, Jaruea, Kevin Jr, Timothy Jr, and Israel, ***Nieces:*** Pam & Family, Kim, and Jade.***Cousins:*** Ike & Debbie Family, Eric & Janette Family, Eric Jr & Family, Little Janette, Paula & Alexis, Carey & Family, James & Family, Marie & Family, PeeWee & Family, Shiela & Family, and Bruce & Family, ***Family Tree History:*** Great Grand-Mother Ada Spencer (RIP), Grand-Mother Pearl Clark (RIP), and Grand-Father Harold Clark (RIP), ***Special People:*** Michele Grobes & family, Bobby & Terry Holiday, Quincy Epps. Vizion Entertainment Group: Tim Sr. Mary, Tim Jr, Israel, and Jade, Strength To Love: Eric & Family, Burlington County Tennis Association (Ike & Family), Sheila Hill & Family, Bruce Johnson & family,Cliff Johnson & family, Cliff Jr & Family, Vincent Moore & family, Greg Tobias & family, Gene & Family, E. Washington Rhodes Basketball Team: Blair Floyd, Marcia & Family, Cobra Jets Baseball Team: Coach Ralph Burrell: Richard Johnson and Family, Reggie Landers & Family, Clark Family, Jones Family, Dave Wright & Nate Wright Family, Arn & Family, Courtland Smith & Family, Booker & Family, ***American Legion Baseball:*** (Oaklane Raidiers), ***AA:*** Baseball (Woodland Giants, and Braves) ***Neighborhood bands:*** Fortified Funk: Brother Kevin, Toby, Ty, and Jeff. Source of Power: Maurice, Puddy, Marty, Calvin, Darren, Nate, Cash, Kim and Ms. Pat. Power House Band, Oasis Band: Vincent, Bey, and Tony, ***Cheyney University Singers and Musicians:*** Tim Sr (Brother), Mike (RIP), Al-Smooth, Greg, Daryl, ,Cecil, Labaron, Barry, Lorette, Reba. **Smooth Jazz:** Daryl, Kevin, Bobby, Alphonso. Blue Magic, ***Empire Records, "Justice League": Rick, Vince, Glen, Stacy and Ron, Philly's BEST Power 99 DC Todd, and WDAS 105.3 Raymond***

Anthony. ***Schools:*** John Greenleaf Whittier Elementary, Booker & Family,E. Washington Rhodes Middle School Teachers: Ms. Rogers, Mr. Albert, Mr. Turner, Ms. Terry, and Mr. McCulley, Murrell Dobbins A.V.T.S (Mr. James Jones, Mrs. Bakum, Mr. Natalone, and Mrs.Reid, Mr. Smith). 307 Academic Social Fellowship: Beta Chapter INC, Cheyney University Department of Engineering, Education, and Industrial Technology / Arts: Mr. James Flemming, Dr. Bagley, Dr. Harris, Dr. Proctor, Dr. Patrick, The College of New Jersey Department of Engineering, Learning Links: Dr. RonTodd, John and Pat Hutchinson, and New Jersey Chamber of Commerce. Reggie & Rosalind, Fran,West Chester University: Educator 500. VillanovaUniversity - Mission to Hubble. Holy FamilyUniversity: Department of Mathematics,

California University: Governs Institutes for Technology, NASA Goddard Space Center, Gear Up, Penn State Robotics Mission to Mars, NASA, Drexel University, MIT, and United States Navy Underwater Robotics Sea Perch, Drexel University RET Department of Nano Sciences and Mechanical Engineering, Dr. Kim, and The Kim Group, Penn State & University of Colorado Space Research and Development: Lisa Brown, Mr. Blocker & Family Darryl (RIP), Kevin and Greg, TEEAP (Technology Engineering Education Association of Pennsylvania), William, Stan, School District of ***Philadelphia Tech-Team:*** Cliff Johnson, Andrew Sears, John Seymour, Daryl Overton, Glen Allen, Keith Washington, Barry Morgan & Family, Joe Huttlin, Tom Goldsmith, Vern, and Jerry. Thurgood Marshall Tech-Team: Mrs. Reyes, Mrs. McClendon, Mr. Hamilton, Ms Glover, Ms. Davis, Mr. Hutty, Mrs. Irean, Jamie, Kenneth, Angel, Dante S, Dante K, Kyle, Randy, Eric, Kim, Daryl, Diamond, Michael, and Mrs Welcome,Philadelphia Secondary Robotics Initiative, BEST (Boasting Engineering Science Technology): University of Pennsylvania Nano Technology FLL (First Lego League), Pennsylvania Gear Up, S. Green & Family Mrs. Velda Morris, Mr. Morris, Daryl Murray ITEEA International Technology Engineering Educator Association,TEEAP (Technology and Engineering Education Association of Pennsylvania, Olney Times Newspaper, Children Design & Engineering: Johnson & Johnson Laboratories Chemistry and Technology, Pat Hutchinson, The College of New Jersey, Glaxo-SmithKline Pharmaceuticals and Robotics, NASA Goddard Space Center, Dow Chemical, Clean Tech (Feed The World), Philadelphia Science and Mathematics Coalition,

Mr. Mel Payne, Don Mckinney (RIP), Philadelphia Educational Fund, Philadelphia Center for Arts & Technologies, Foundations Inc, The Montessori School, Laura Stulb, Janet Easlea, Seema Khetarpal,

Dottie Baumgarten, TMS Administrators, Board Members, Teachers, Students, Staff Members, and Parents, Dr. Hevrin &Family, Veronica, Miguel, and Nina, Marie Lambert & Family, Kevin Jones & Family, India & Barry Family, Pastor & Mrs. Hasty, and Bishop & Mrs. Stewart.

Additional Special Thanks:

Education:

Murrell Dobbins AVT High School
SDP (Philadelphia Tech-Team)
Cheyney University (Department of Engineering, Industrial Arts & Technology, and Education)
The College of New Jersey (Project Update)
The College of New Jersey (Children Designing and Engineering)
The College of New Jersey (CAD / Pro Desktop)
Drexel University (Nano Sciences, Mechanical Engineering, and Underwater Robotics)
University of Pennsylvania (Department of Engineering - Robotics)
Penn State (Nasa Space Exploration)
Villanova University (Computer Aided Design)
Auburn University (Hubble Space Telescope Research - Robotics)
Holy Family University (Mathematical Concepts)

Professional Organizations

ITEEA (International Technology and Engineering for Educators Association)
TEAP (Technology Engineering Education Association of Pennsylvania)
Little Kids Rock / Modern Band Summit
Educator 500 (West Chester University)
Governors Institute for Technology (California University of PA)

International Companies:

Nasa (Goddard and Johnson Space Center)
Johnson & Johnson Laboratories
Glaxo Smith Kline
Six Flags
Legos Corporation
Knex Corporation
Sam Ash Music
Apple Computers
Makerbot Education (3D Modeling / Printing)
Play-Doh

Schools:

Frankford High School
Thurgood Marshall K-8 School
The Montessori School

STEM Educational Enrichment Services LLC.

The Montessori School
PCAT (Philadelphia Center for Art and Technology)
Dow Chemicals
Glaxo Smith Kline
Franklin Institute
Philadelphia Educational Fund

Made in United States
Troutdale, OR
11/03/2026

56847406R00132